AF480785

Praise for *Love Conquers Fear*

"With the scientific world converging on a post-materialist reality in which mind plays a powerful role in the emergence of physical reality, unification of science and spirituality provides a natural path forward to remove fear from our world. In this extraordinary book, Brett offers a focused, yet encyclopedic review of how humanity has arrived at our current state over thousands of years. The second half of *Love Conquers Fear* is devoted to a powerful vision of how to blend collective (Eastern) and individual (Western) cultural strengths to empower a heart-dominant humanity and achieve abundance for all. His deep knowledge of the Superfecta (AI, quantum computing, robotics and brain-computer interfaces) provides the engine to these extraordinary possibilities. This masterpiece should be required reading for all modern humans. Most highly recommended!"

> – **Eben Alexander, MD, former Harvard Neurosurgeon and author of *Proof of Heaven*, *The Map of Heaven*, and *Living in a Mindful Universe***

"Whatever your concerns or enthusiasms about AI, consciousness, or religion, this carefully researched and erudite book has something to offer you. Brett Hurt resists the pull toward certainty—toward applying yesterday's answers to today's and tomorrow's questions—and the alternative he proposes is worth very serious attention: why choose fear when the alternative opens up a better world for all of us."

> – **Ellen Langer, Professor of Psychology at Harvard, "the mother of mindfulness", and bestselling author of *The Mindful Body* and many other amazing books on the power of mindfulness**

"*Love Conquers Fear* is a masterpiece; and one that only Brett Hurt could have written. No one else on the planet could so authoritatively synthesize topics as seemingly unrelated as artificial intelligence, transcendent spirituality, material abundance, and human consciousness. Ideas are the most powerful forces in the universe, and this book has big ones on almost every page. I doubt that anyone can read this book and afterwards not see the world—and their place in it—differently."

"We are entering a decade where technology will completely reshape every aspect of human life. Brett makes the case that the real question isn't what we build, it's who we become while building it. In *Love Conquers Fear*, he connects dots that are too often kept separate: advanced technology, human consciousness, and the responsibility of those who lead humanity forward.

The core message is profound, yet simple—fear leads to scarcity, control, and short-term decisions that self-sabotage; Love leads to expansion, alignment, and a version of humanity beyond anything we can dream of. While the framing is simple, the implications are enormous, especially as AI, quantum computing, robotics, and brain-computer interfaces accelerate.

Brett brings enormous credibility and courage to the questions of our time. He understands these technologies at a deep level, and fully embraces the hardest questions: how do we ensure these technologies serve humanity, not just greed and power? This book challenges leaders to think bigger, act with more intention, and take responsibility for the direction we are heading, while acknowledging every one of us has a role to play.

If we get this right, the upside is more extraordinary than anything humanity has ever experienced. If we don't, the cost is potentially unfathomable. *Love Conquers Fear* helps you see that clearly—and more importantly, helps you choose."

"Brett has written a rare kind of book that feels both deeply personal and wildly expansive. *Love Conquers Fear* draws on decades of his extraordinary experience and tireless research to offer grounded optimism that feels like oxygen. At a time when so much of the world is driven by fear and division, Brett reminds us that we have agency, that another path is possible, and that the way forward begins within. His ability to blend brilliant insights from technology, consciousness, and ancient wisdom points us not just toward a more unified future, but back to our own hearts... and I can think of no greater gift to humanity right now.

As someone devoted to self-inquiry and the practice of returning to truth, I experienced *Love Conquers Fear* as a kind of companion on that path that meets this moment with profound clarity and depth. I was struck by how seamlessly it weaves the outer world of rapid technological change with the inner journey back to the Self. In these pages, Brett is also building bridges, softening the edges of our most entrenched dualities and revealing a deeper coherence beneath them. By the time I reached the final pages, I felt more open, more at ease, and more connected to that quiet place within that knows. This is a book that inspires hope by helping you recognize it in yourself.

If destruction is one of the defining forces of our time, then *Love Conquers Fear* stands is a rebellious act of creation that reminds us of what is still possible when we choose to build, to heal, and to awaken. It is an offering that carries the potential to ripple outward in ways we may never fully see. This is the book the world needs right now."

- **Michael Collins, Director/Producer/Writer of Thoughtful Robot Productions**

"I believe now more than ever that raising human consciousness across our collective is by far the most leveraged thing we can do to realize our species' infinite potential, now all dramatically multiplied by AI. *Love Conquers Fear* is a rare work that bridges the worlds of technology and spirit with the drive of a serial entrepreneur and the heart of a genuine seeker. In this lifetime, I've never seen someone come so far so fast on their Truth-Seeking journey as Brett. The same rigor, passion, and heart he brought to his soul quest comes across throughout this book. This book is a loving call to action for us as a species to overcome the shackles of fear and the illusion of separation and to boldly venture into our birthright: realizing our deepest, truest, infinite potential. This mirrors my own thesis for several years now—that what unites us is far greater than what is perceived to divide us."

"*Love Conquers Fear* is a deeply researched and extremely thoughtful book with a critically important message: unless we continually struggle to realize our best values, the technological revolutions offering so much promise will instead do unnecessary harm. The book is a call to action. Shame on us if we don't hear it."

"There is no business to be done on a dead planet and mass unemployment. Positive sum thinking must win over zero sum thinking. Love must win over fear. *Love Conquers Fear* is a must-read book from my friend Brett Hurt."

"*Love Conquers Fear* hit me with the force of a tsunami. It's far more than a treatise on technology: it is a profound guide to reclaiming our souls in a world that feels increasingly fragmented. Reading this book did not just change my perspective on the future. It fundamentally altered my understanding of the very nature of reality.

Brett masterfully deconstructs the materialist illusion that has long kept us in a state of spiritual sleepwalking. He invites us to recognize that consciousness is not an accidental byproduct of biology. Instead, it is fundamental: the very substrate of the universe. This insight is a beacon of light in the current chaos. It reminds us that we are not merely data points in an algorithm but vital nodes in a cosmic lattice of interconnected love.

At this critical time in our history, AI is forcing us to ask the ultimate question: What does it mean to be human? Brett frames the 'Superfecta' of AI, quantum computing, robotics, and brain-computer interfaces not as tools of replacement but as mirrors of our own potential. He challenges us to move from Homo sapiens to "Homo techne", using our god-like technology to amplify our empathy and wisdom rather than our weapons.

This book is a manifesto for Team Humanity to pass its most crucial exam. It moved me to realize that the most important code we will ever rewrite is our own. For anyone feeling the disorientation of this awakening, *Love Conquers Fear* is the map you have been waiting for. It is time to let the heart finally conquer the head."

- Fernando Espuelas, CEO of Vera AI

"For nearly twenty years, I've had the privilege of working alongside Brett as he built companies and communities driven by a genuine desire to make the world better. In *Love Conquers Fear*, Brett brings together technology, philosophy, and spirituality to challenge us with one of the defining questions of our time: will we allow fear to divide us, or will we choose love to guide us? At a time when our world feels increasingly divided, For You leaders look for the threads that unite people and move humanity forward. Love does that—and Brett calls us to lead with it. For me, the deepest source of that kind of love comes from my faith, and Brett's message reminds us why it matters. I strongly recommend you read this book, then decide for yourself how you will lead going forward."

- Kirk Dando, CEO Coach, Investor, and Host of the *For You Leaders* Podcast

Read more praise for:

LOVE CONQUERS FEAR

... inside the back cover.

LOVE
CONQUERS
FEAR

An AI-generated *frontispiece* from a prompt written by
Claude and executed by ChatGPT.

A *frontispiece*—or simply *frontis*—is an illustration or
plate inserted immediately in front of the title page,
with the illustration facing the title page.

Also by Brett Alexander Hurt

*The Entrepreneur's Essentials: Lessons for
Startup and Leadership Excellence*

The Lattice: A Novella

LOVE CONQUERS FEAR

Humanity, AI, and the Age of Abundance for All

by
BRETT ALEXANDER HURT

LOVE CONQUERS FEAR: Humanity, AI, and the Age of Abundance for All

Cover art by Jacob Salamon.
Book design by Ryan Simanek.

ISBN: 979-8-90387-000-4 (hardcover edition)
       979-8-90387-001-1 (paperback edition)
       979-8-90387-002-8 (eBook edition)
       979-8-90387-003-5 (audiobook edition)

Published by Do More Good® Publishing

Printed in the United States of America

Dedication

To Team Humanity,
which will bring us into the
Age of Abundance for All.

As love conquers fear,
the future that awaits us is unity
across this beautiful planet Earth
and the cosmos beyond …

Contents

List of Illustrations

Foreword

Remember, Remember, Remember

Love Conquers Fear: Humanity, AI, and the Age of Abundance for All (LCF) is about something bigger than any individual, company, industry, or nation—it's about the future of humanity and the coming age of transformation brought about by AI. This book is divinely inspired at a time when many people are experiencing cultural, social, technical, or spiritual disruptions, prompting us to examine our identity. *LCF* asks us to consider what we stand for, where the line is too far, and, in so doing, to choose between whom or what we love more. It is a manifesto, a bold moral call to action that inspires leaders at all levels of society to examine their strongest motivational forces: love and fear.

The books referenced, and the associated bibliography, constitute a cornucopia of philosophy, leadership, science, technology, and spiritual teachings that would rival the library of Socrates. *LCF* deserves deep contemplation and analysis. It contrasts an Old World Mosaic, motivated by fear, scarcity, and greed to control and manipulate, versus a New World Mosaic, motivated by love, abundance, health, wellness, and sustainability, while applying a capitalistic approach that balances all stakeholders, not just shareholders. It emphasizes that time is of the essence for shifting our collective rudder and embracing the teachings of all the great masters.

I know Brett Hurt deeply, and he is a profound human being

with a huge heart. He answered the calling and chose love over fear, others over self, nature over attachments, and satisfaction over gratification. We should all be grateful for his courage, dedication, and perseverance. He is also a rarity in this field—not (previously) one of the countless scholars, philosophers, spiritual teachers, or scientists who, for decades, have written about the coming new age for humanity, but rather a tech titan. He is the founder of multiple businesses, who began coding at age seven. Whether you believe in God, Source, the One Mind, or Creator, the Infinite either has a good sense of humor or a secret plan to use technology in a masterful, final, quantum act. *LCF* is divided into three sections, with recurring themes: Definition, Application, and Transformation. It begins appropriately by discussing love and fear, and I want to share my personal experiences of these emotions.

Looking back on life, we often discover a turning point, a moment that shifts one's trajectory, shatters paradigms, and gives birth to manifestos. In my case, it happened at the pinnacle of my professional career when I was the Chief Executive Officer of EY Management Consulting in Singapore: I broke my neck, and my spirit left my body.

At the time, I was successful by worldly standards. I thought I knew who I was, where I came from, and why I did what I did. I thought that everything I did was an independent event. More importantly, I thought I was happy. I *certainly* thought I was in control of my life and the thousands of employees I managed. I enjoyed leading global transformations. I had three beautiful, healthy, and smart children, and a faithful and devoted wife. Not bad for a kid who grew up with hand-me-down clothes and an alcoholic father who beat me.

In retrospect, before a fateful Tuesday that would change my life forever, I was pretending, playing a game, hiding behind a mask, like other CEOs I know. Yes, I was captain of my own ship, but imprisoned in an ocean of prevailing social, economic, political, and ecological winds and currents that blew me where they directed. Unbeknownst to me, everything I thought I knew was about to be annihilated. Before my Tuesday wake-up call, I was ignorant of the unseen forces that influence everything, everyone, and everywhere, which is shared in *LCF*. I was also unaware of my inner demons, childhood wounds that were anchors, holding me back from being an authentic person. Although I may have been the only visible one holding the rudder, just as the wind, tide, and currents can blow a ship onto the rocks, so too are there powers that send mortals into the forest searching for what they know not. These moments are not planned. They arrive and change lives, just as this manifesto does.

On Tuesday morning, January 18, 2000, I did not arrive at the office. Like the proverbial frog sitting in a pot of room-temperature water and being boiled alive as the temperature was raised one degree an hour, a slow change was burning inside. The nerves in my spine were being ground away, bit by microscopic bit, from a couple of mortars called vertebrae. After all, I lost the sponges called discs fifteen years earlier, on the U.S.S. Tecumseh in 1985, when I was a nuclear submarine officer. The ship was in dry dock in Newport News Shipyard, sitting in a cradle as shipyard workers cut giant thirty-foot holes in the hull so they could remove all the engines, turbines, and pumps. Around me were ten-foot sparks streaking by. The burnt-butter smell, the steel being sliced away, was strangely appealing. Stadium spotlights hundreds of feet in the air, peering down on the night-shift workers as cranes circled between the glow and fog. Yet I had to stay focused; I was supervising the refueling of the nuclear reactor. A crane started swinging a load and struck me in the head, snapping my neck back, crushing three vertebrae. My recovery took months, and I was pain-free for almost fifteen years. By the summer of 1999, that all changed as my neck deteriorated.

I woke up and got out of bed. I walked to the bathroom and turned on the shower to let the water heat up. I raised my arms for a morning stretch. That is when it happened. I screamed as I felt a fifty-pound, six-inch axe trying to slice me in half. It sliced between my shoulder blades as my collapsing, twisted vertebrae crushed and ground away on nerves that ran up and down my spine. My face contorted, and the screams bounced off the marble and hammered my eardrums. I staggered on one foot and fell against the wall as I tried to crawl my way out of the bathroom. I raised my arms over my head and tried to pull the axe out as my vision narrowed, a black tunnel approached, and stars flew. The reaching created another tectonic explosion inside, and I buckled forward, then snapped up as if I had a swift kick in the jaw. My entire body began to shake uncontrollably as I was being electrified by my own power plant. My wife, Marianne, came running to my rescue, just as my body was falling, but I was no longer in it.

My vantage point was existential, I was out of my body, looking down from a height of eight feet, as Marianne tried to catch my body as I—perhaps I should say it—collapsed. Its arms were limp, its head drooped, and its legs buckled. The head hit the corner of the bedroom dresser as the body flopped down to the floor like a puppet whose strings were severed.

I was sucked back into a blue, pulsating tunnel like Jody Foster in *Contact* or Rey in *The Force Awakens*. I turned left, then right, ninety degrees down, spinning and swirling. There were beautiful streaks of white sparklers twirling and twisting inside the blue tunnel. I was inside a fireworks show or aboard the USS Enterprise at warp speed. How long was irrelevant; time was non-existent, as was my body. As I exited the tunnel, I was flung into dark space and approached a gateway. Over two decades later, and I can still see and feel these moments like they were yesterday. The gateway was floating in deep space, the most beautiful thing I had ever seen or would ever see, twirling, glowing, and vibrating. It looked alive. There were seven triangles inside triangles, each a different color; collectively all the colors of the rainbow, pulsing with such vibrancy words cannot adequately express their beauty. If I could have cried, I would have, but without a body, there were no tears. The series of triangular gateways opened one by one as I passed through. After the final lock opened, my consciousness was flying through space and time. Not only was there no pain or fear, but it was also peaceful. Words cannot explain the enhanced sensations, a fully lived moment, and the feeling of being connected to everything, everyone, and everywhere. It took years of meditation, yoga, and mindfulness practices to achieve this transcendental state once again.

Two weeks later, after a double surgery to remove bone from my hip and use it to repair my neck, I was removed from a morphine induced state, recovering at home, immobilized in braces, and unable to walk or barely talk. As I lay in bed, staring at the swirls in the plaster ceiling above me, I began to cry. I led mergers while my family climbed the Great Wall, and I designed outsourcing projects while they went to Sydney and played with Koala bears. I lived to work. Why? I shouted a silent prayer in my head, demanding God to tell me, "Why did you do this to me?" In less than one minute, my three-year-old son jiggled open the door and came into the room. He danced in circles and waved his arms in the air as he shouted "Wemember, wemember." He could not pronounce an R. He left as fast as he entered. It took me six months to learn how to walk, talk, and chew gum, and I was never the same again. I was compelled to figure out why I was successful but unfulfilled.

Over the next ten years, I began a quest to answer the great questions: who am I, where did I come from, why am I here, and what am I supposed to remember? I studied Reiki, craniosacral therapy, energy healing, massage, chiropractic care, kriya yoga, hypnosis, regression therapy, and more. I learned these modalities out of a necessity to heal trauma that I did not know I had. Collectively, they exposed a virus sleeping in my mental software—feeling unloved and unlovable.

Stories such as mine are not hard to find. Whether another business professional, medical doctor, or Nobel Laureate, they share the same experiences and shattered paradigms. In my case, the words of a three-year-old were understood: Remember who I am, remember where I came from, and remember why I am here. I began to wonder: When did I stop feeling with the heart behind the heart? When did I stop seeing with the eye behind the eye? When did I stop hearing with the ear behind the ear? And what role did fear play?

There are moments in a life when something we face does not feel new but remembered. Not remembered by the mind, but by something deeper—something that recognizes truth before it can be explained. A resonance. A soft internal nod. A subtle loosening of a long-held belief. This book arrived for me that way, and when it did, I did not skim it; I read it slowly and absorbed it. I again *remembered* all the misconceptions and illusions—the social, political, economic, ecological, corporate maya—the source of my confusion, suffering, and attachments, so eloquently and holistically named in this book. I *remembered* the U.S. Naval Academy nuclear submarine officer whose dream of retiring as an admiral ended after realizing I could not push the button to launch the nuclear missiles. Back then, humans had to push the button and had time to think about it. Let us hope that AI is never the decision maker. I *remembered* the Exxon executive who volunteered to supervise cleaning up 11 million gallons of crude oil spilled in Prince William Sound from the Exxon Valdez, who one day marveled at the majestic beauty of whales and glaciers, and the next day cried, horrified, at the sight of 500,000 dead birds and fifty thousand dead otters. Today, I wonder why we are not tapping into more of the most powerful source of energy, the sun, as Brett suggests. Perhaps AI can help us get there. I *remembered* the EY CEO who designed and led the globalization efforts, saw the countless rounds of layoffs and the harm they caused families, and resigned in protest. I was reminded that Adam Smith said capitalism should be guided by trust, moral restraint, justice, and fair competition, and that Benjamin Franklin argued it is the proper balance between commerce and philanthropy. And I *remembered* the Hurricane Harvey crisis manager in Houston, Texas, during the nation's worst hurricane disaster, after five feet of water damaged more than 200,000 homes and businesses. Because of Brett and this book, I now wonder how AI in the Age of Abundance could help with sustainable living. *LCF* addresses all these topics and more.

One day, the spirit calls, and when it does, resistance is futile. Trust that its arrival is no accident. It teases us to choose between whom or what we love more. It whispers in our ear while our head is on the pillow, "Is silence still an option?" It stops the car in the middle of a road trip and taunts, "Is the story you are telling yourself still working for you? How many signposts have you missed?" When love conquers fear, you will find your answers.

From every realm, be it corporate, political, social, or spiritual, there is an epidemic of leaders frozen in fear and stuck. Many are like I was, feeling unloved and unlovable. They are confused, as was I, by the timeless battle between their ego and soul. The head tells them to please the status quo, be that Wall Street, Main Street, or Pennsylvania Avenue. The purpose of a company is to make money. The ego is unconcerned with launching missiles, spilling oil, polluting the air or food, or building houses in a floodplain. It is a business, and neither animals, the planet, nor families are included as performance measures in the compensation plan. Not yet, anyway. There are no tears. Men do not cry. Their soul, however, tells them to protect and care for their people, the air we breathe, the water we drink, and their children born into a new paradigm who care not for the artificial boundaries that separate us by race, gender, sexual orientation, or national origin.

But who can free themselves from the prison of golden handcuffs or personal trauma? Who dares become the authentic, congruent, and inspiring leader the world desperately needs, that *Love Conquers Fear* so tenderly calls? Had I not been in this situation myself as a Chief Executive Officer, had I not had a near-death experience, I would not have understood the dilemma, nor would I have had the profound experiences to understand, let alone embrace or love, *The Way* of this book. Brett's wisdom and its application in these times constitute a tender knock on the door for every leader, "Hello, are you out there, do you still feel and hear me?" *Love Conquers Fear* is the title and the singular most crucial recurring motif, and I am reminded of it nearly every waking moment when social media attempts to fill me with fear and division. Without my spiritual transformation and healing, I would have been unable to recognize and shift the outdated paradigms or ignore the clickbait.

At a time in our collective history when it is impossible to go a single day without facing divisiveness and hatred, Brett reminds us to remember that we do not just have a soul; we are a soul having a physical experience. *Love Conquers Fear* is not just about technology, AI, economics, geopolitics, or even the future. It is about choice. The kind we make quietly, often unconsciously, in every waking decision. The choice between contraction and expansion. Between protecting an identity and serving something larger. Between fear, which isolates, and love, which integrates. It is not presented as an argument to be evaluated, but as a mirror held at just the right angle for us to reflect. It did not tell me what to think. It asked me what I was afraid to feel. It did not offer certainty. It invited responsibility. And it did not demand belief—only honesty. If one looks to change the world, it begins by looking at oneself.

As Gandhi once said, *"Our beliefs become thoughts, which become words that become actions, and form behaviors that become our destiny."* Buddha said, *"With our thoughts, make the world."* Einstein called this quantum entanglement "spooky action at a distance"—our thoughts are floating around at a quantum level and coagulate, in a sense, to create physical reality. Henry Ford once said, *"If you think you can or think you can't, you are correct."* A spiritual master, a Nobel laureate in physics, an industrial titan, and a tech titan; all sharing the same wisdom. Ponder on this if you care.

Viktor Frankl once asked, the ultimate question is not what we expect from life, but to what—or to whom—we are responsible. This book does not answer that question; it cannot be outsourced. Nor does it claim to have all the answers. What it does—gently but persistently—is refuse to let us avoid it.

Brett Hurt reminds us to lead with love and to balance purpose and profit, for our collective survival depends upon it. When love conquers fear, we all rise. It is in this vein that the Foreword was written. Each of us must choose. This book reminds us of something we have forgotten and asks us all to *remember, remember, remember: Remember* who you are beneath the roles and the noise—you are not a title; *Remember* what matters when the metrics fall away—money does not bring happiness; *Remember* that love is not the absence of fear—but the courage to move through it in service of something greater.

If something in you is ready, I encourage you to continue the journey, for it is an experience, not a destination. Consider this an invitation. If you are ready to accept it, turn the page.

JD Messinger
Executive Director, Do More Good® Movement
Author of *11 Days in May: The Conversation That Will Change Your Life* Five-time award winner and #1 Amazon Best Seller in Personal Development
January 24, 2026

Acknowledgments

Love Conquers Fear, both this book and the podcast, are a labor of love. This is my soul song, it's my love letter to humanity. So, my choice of partners really mattered.

There is simply no way I could have written a book of this scope without the partnership of David Judson. We began working together back in 2020 when he was the co-founder and Editor In Chief of Urbānitūs. David and I immediately clicked, and we partnered to write some of my favorite articles through the fog of the COVID-19 pandemic and Austin's acceleration. That work continued into my first book, *The Entrepreneur's Essentials*, which had begun as an online-only, solo project to codify my entrepreneurial learnings. David helped me make it much better as we expanded on it in print. It was only natural to work with him on *Love Conquers Fear*, and it has been a real joy and honor to do so. He and I are jointly dedicated to helping humanity have a heart shift during this fateful time, and it is clear to me both in this lifetime and in my soul that he is a very important guide. I chuckled as I looked up his new title on LinkedIn while writing these Acknowledgments. Wayfinder. Yep, that he is!

Lincoln Brown, Patrick Gentempo, Jason Karp, JP Newman, Mark Gober, Niraj Mehta, Gaby Poler-Buzali, and Eric Wolf have been tremendous guides and close friends along the way in this journey. I'm very grateful to them.

John Mackey, Piraye Beim, Sol Rashidi, William Steele, and the whole crew I went to Necker Island with in March of 2025, thanks to

Alexander McCobin, were foundational in this journey.

All of the guests that have been on the *Love Conquers Fear* podcast were essential learning nodes and really showed their heart in person and on camera. And there's no way the podcast would have kept on rolling out so smoothly without the help of Jacob Salamon and the team he assembled. Thanks to Stacey Rae and her team for initially helping to start the engine.

Jay Wilkinson, JD Messinger, Phil Whitmarsh, Jorge Armando Cazares, Lori Martinsek, and the entire Do More Good® Publishing team have been great partners. I wanted a publishing house that really got me, and that they did. I wouldn't have met Jay and his team without the grace of Randy Cohen inviting me to give that speech at MIT to his fifty CEO buddies.

All of the authors and teachers that have helped me on this journey, including those that have become such good friends, like Mark Gober, Chris Bache, and Eben Alexander, have given me the education of a lifetime... this lifetime, that is!

My soulmate, Debra, of thirty years of marriage and our children, Rachel and Yuzu, have been a constant source of inspiration. My sister, Brandi, and niece, Chantel, have been as well. I love you with all of my heart. Thanks to everyone that has worked with and believed in me as an entrepreneur and CEO, from co-founders and team members to investors and Board members. And thank you to all of those that have allowed Debra and me to invest in their startups and venture capital funds. The challenge of creating something new is both daunting and exciting. I've seen firsthand how love can conquer fear over and over again.

Thanks to Rabbi Neil Blumofe, Guruji, Swamiji Parthasarathy, Paige Britt, and other spiritual teachers that I count as friends for helping me grapple with the nature of our reality.

And, finally, none of this would have been possible without the divine. Thank you for choosing me as one of your vessels on this mission to help humanity reach its crescendo as love conquers fear. And thank you for putting that divine spark in us all. Your love for us, all of us, knows no bounds.

This book has certainly been an ambitious labor of love, and I'm sending you much love and gratitude for diving into it. This past year, 2025, was one of the most remarkable years of my life.

My soul quest throughout 2025 and beyond began in December of 2024 in Kerala, India. Kerala is a mystical place, where you traverse what the locals call the backwaters on unique houseboats and experience the magic of Ayurvedic treatments at the hotels. I've always loved India with the deep feeling that in one of my past lives I was an Indian. Loving the flavorful food as a vegetarian and having been drawn to the deeply spiritual cultural roots of Vedanta, a profound Indian philosophy on living a happy life that the Young Presidents' Organization (YPO) introduced me to, India felt like a home away from home to me, even with its obvious chaos of traffic and poverty.

What had brought me to Kerala was a beautiful ceremony, the first Indian wedding I've ever experienced. When our daughter, Rachel, had Rishi, one of her Tulane University friends and fellow classmates, over at our home during Tulane's spring break in 2024, Rishi invited us to his older sister's wedding in Mumbai. It was an immediate yes for both my wife Debra and I, and we went all in. Rachel and I drove up from Austin to Dallas for Thanksgiving, and we had gotten tailored clothes at a traditional Indian clothing store while we were there. Ultimately, Debra and our youngest child, Yuzu, decided to not make this voyage to India at the end of the year, opting for a quieter close to 2024. The wedding was a three-day event and was absolutely stunning in its beauty. It was not just our first Indian wedding; it was the first wedding I had ever been to with over 1,000 people!

I was relaxing from the hustle of the wedding with Rachel in Kerala. She stayed for most of the time at the Ayurvedic spa resort I had chosen and then flew back a few days early to be at home in Austin for New Year's Eve. So, here I was thinking what to do next besides experiencing more Ayurvedic treatments.

I remembered that my good friend, Jason Karp of Hu Chocolate and hedge-fund fame, had recommended the documentary *Inner Worlds, Outer Worlds* to me as a good unifying piece on spirituality and science. I watched it for free on YouTube. Although it was just two hours long, the film resonated on such a deep level for me that it took me six hours to work through it. That year, I'd had several mystical experiences with psychedelics, which included meeting God during one of my journeys. The feeling of meeting God in a journey, or whatever you choose to call that force that created the universe with the mystery of the Big Bang, was as profound as could be. In the presence of God I felt completely unconditional love and acceptance, a warmth surrounding me that I cannot quite explain but felt like the ultimate bliss and, to use Michael Pollan's moving word from his brilliant book and documentary, *How to Change Your Mind*, it was *ineffable*. In Chapter 13, "The Inner Journey," I'll share some of my transcripts from a later psychedelic journey where I met God again, and the experience was as deeply personal and sacred to me as can be. It was the journey that directly led to the insight that love must conquer fear for humanity to make it to the Age of Abundance for All. I knew helping humanity overcome fear was a divine directive that I needed to act on to help secure humanity's future, and I started working on this book and the podcast shortly after.

Since watching *Inner Worlds, Outer Worlds*, I've read fifty-five books on what I'll call my soul quest following that fateful trip to Kerala. I cover a distillation of the best of these in Chapter 12, "What Is Our

Reality?" I've also distilled my learnings in a science fiction novella named *The Lattice*, which paints a positive picture for humanity to make it to the Age of Abundance for All. The protagonist in *The Lattice* is named Dr. Alexander Bliss, who is a prominent physicist expert in quantum chromodynamics, the type of subject matter that scientists at the *Conseil Européen pour la Recherche Nucléaire* (CERN), which translates to the European Council for Nuclear Research, obsess about. It has always been our nature to question our reality, and CERN has some of the best scientific tools to do so. Quantum chromodynamics is the theory that describes the strong nuclear force that binds quarks and gluons into protons, neutrons, and all atomic nuclei. It's a core part of the Standard Model of particle physics—a theory describing the basic building blocks of matter and how they interact. CERN is where the Higgs boson was discovered using CERN's Large Hadron Collider. It was the final missing piece of the Standard Model of particle physics. The discovery of the Higgs boson confirmed the existence of the Higgs field, a pervasive field that gives mass to other fundamental particles.

That field is what *Inner Worlds, Outer Worlds* is all about. The documentarian, a Canadian filmmaker named Daniel Schmidt, turned to deeply spiritual texts to show how this vibratory, unified field was already known thousands of years ago. Schmidt founded the Awaken the World Initiative, a non-profit project focused on producing and distributing spiritual and consciousness-oriented media and resources for free in dozens of languages. In *The Lattice*, Dr. Alexander Bliss not only proves that we are accessing that field, which he calls the lattice, in psychedelic journeys but also in meditations, near-death experiences (NDEs), and through deeply connected spiritual masters. He shows that the lattice is deeply rooted in Oneness and love, which proves the existence of a divine field that some, like me, choose to call God. Humanity unites and we have an unprecedented technological takeoff from the year 2031 to 2085.

Of course, this is the future that I deeply want for humanity. And I know from my journeys, prayers, and meditations that is the future that God wants too. If the word *God* triggers you, I understand. Throughout history, we have had religious persecution and cancellation.

Galileo was famously cancelled by the Catholic Church for urging them to look through his telescope to show that the universe did not revolve around Earth. The Roman Inquisition tried him in 1633, found him "vehemently suspect of heresy," and sentenced him to house arrest for the rest of his life. And his 1632 book, *Dialogue Concerning the Two Chief World Systems*, which strongly defended heliocentrism, was banned. My view of God is not one where I'm personifying Him/Her/It. I don't profess to know what God is. But I do know that we can glimpse the Oneness of all things rooted deeply in love within a psychedelic journey or a deep meditation. The number of conversations that I've had throughout 2025 on this subject have opened up my mind and heart to the truth. I've found out now that twelve of my friends, some of whom I've known for decades, have had NDEs where they've seen the Oneness rooted deeply in love too. Before 2025, do you know how many friends told me about their NDEs? Zero. Like saying the word *God* or discussing psychedelics, talking about NDEs is taboo. Meditations are ancient and what the Buddhists were practicing thousands of years ago to create internal bliss. As Christianity would say, "The kingdom of God is found within."

As a successful capitalist, I realize that on the surface I may be seen as an unusual type of vessel to write this book. That's also the point. I've been shown in my journeys that I need to do my best to help unify science and spirituality; that the dogmatic separation between the two has led to a tremendous amount of suffering. In his book *The Seat of the Soul,* Gary Zukav posits that the five-sensory human (what I'll call a materialist), is the root of our suffering. Without knowledge of the Self or the soul, we go through the motions in life with this belief that nothing matters unless we ascribe meaning to it. The materialist believes that consciousness just magically evolved through natural selection and upgraded brain chemistry, mysteriously materializing from matter in a cold and lifeless universe, and that Fermi's Paradox shows there is nothing besides us out there or we would have already met other intelligent life. And that maybe this whole reality we find ourselves in is just a simulation and in videogame speak many of us are Non-Player Characters (NPCs). I wrote a piece on simulation theory addressing my views on that with my good friend and fellow author and entrepreneur, Byron

Reese, titled "No, We Don't Live in a Simulation."

The reason I'm a successful capitalist is that I was born with the right passion at the right moment in history. I had an incredible mom in Brenda Lorelle Hurt. She saw that I was interested in technology since I was age four. I was not content with just playing Atari's Pong video game console (the only game you could play on it) but also wanted to learn how it worked. I was hooked. At age seven, she bought my first computer, an Atari 2600 with the BASIC cartridge, thinking it would make me more interested in mathematics as a bonus. It certainly did, and she learned how to program with me. It was all I wanted to do and I did it for over forty hours a week from ages seven through twenty-two! Growing up in Texas, my mom defended me the entire time from the many voices, both family and teachers, that were telling her that she was raising me wrong and my life would be ruined as a result. I can still hear them saying, "But, Brenda, all he does is work on the computer; he has no balance in his life whatsoever." I truly have no hatred in my heart for hearing any of that, I've fully forgiven. My fascination with computing led to me starting six companies, including one that had a unicorn IPO, and being an investor alongside my wife, Debra, in 150 startups and counting (twelve of them unicorns now), plus fifty venture capital funds and one cryptocurrency fund.

Once I read the book *The Celestine Prophecy*, I started to see synchronicities everywhere throughout 2025. I cold-reached out to Mark Gober, author of two of my favorite books on this soul quest (more on those later), and he immediately felt the energy, agreeing to come to Austin to speak with me at two events, and then moved to Austin (I told him he would and am very thankful he agreed to write the Afterword to this book.) Mark and I attended a taping of the music TV show *Austin City Limits*. Before seeing the artist, Cam, we had an incredible dinner discussion. Cam said that she was going to do something she never did. She played her entire album written during COVID-19. It mirrored our dinner conversation almost exactly. Neither one of us had ever heard her music and we kept looking at each other with surprise, knowing the divine is playful. We covered this moment on the *Love Conquers Fear* podcast, episode 2. The mentalist Oz Pearlman was handing out a

Frisbee at the 2025 TED conference in Vancouver to Dylan Taylor, one of six other Henry Crown Fellows at the conference in an audience of around 1,500, who I know quite well. He threw it and I knew it would come to me, even as it arced in the air. You can see what happened next on this TED video. When you watch that, do I appear to be surprised that it was coming my way? I write about the mind-blowing synchronicity that I experienced in the Grand Canyon in Chapter 13, "The Inner Journey." As I write this Introduction, I'm on the last day of a Siddha Maha meditation retreat with Guruji thanks to my good friend and fellow capitalist Niraj Mehta, and that has been profound and helped me infuse even more love into this writing. This retreat is the first time in my life that I've meditated for five days in a row, and I already know that my 2026 theme for the year is going to be *embodiment*. My 2025 theme was *the inner journey*, as I wrote in my annual New Year's letter. And the list of synchronicities goes on and on. I see them typically two to three times *per day*.

There were many signs along the way that I should write this book. Throughout 2025 I felt the presence of the divine urging me on. In March I went to Necker Island for the first time with John Mackey. We were visiting Alexander McCobin, founder and CEO of Liberty Ventures, who had set up this beautiful gathering with amazingly aligned conscious capitalists. When we pulled up, I saw a man in swim shorts, no shirt on, running towards us on the dock. I thought to myself, *Wow, that's really cool that the team that works here on Necker is so hospitable.* As the man came into focus, I realized, *Wait, that's Richard Branson, the owner of this island and the founder of Virgin!* We had a blast with him for much of our time there. He was the one encouraging me to read *The Alchemist*, which you'll hear more about in later chapters. Necker was the first time that I gave a talk on the hard problem of consciousness in the exponential age of AI, quantum computing, and robotics. I later added brain-computer interfaces to that list of exponentials, and I term these exponential technologies the *Superfecta* in this book. I was swarmed after that talk, with many thanking me for being "so brave to talk about spirituality," which wasn't what they expected, and I've included a link to it here. The next month, I was speaking to a gathering of fifty CEOs at MIT;

same topic, but a much larger format. This was a group that had been gathering for over twenty years. My talk was so well received there that I spent hours after talking to them. I got the same reaction, thanking me for bringing science and spirituality together during this fateful time for humanity's future. I'll also include a link to that here. ❦ Then there was the TED conference, which is a favorite and I've attended eighteen times to date. This year was the first that Chris Anderson pulled me into a special meeting with a group of "AI experts" to help him prepare for his interview with Sam Altman. Chris did an absolutely fantastic job in that interview. ❦ I made the point in our group that consciousness should be front and center. Chris agreed. Without consciousness, which is the basis of everything we cherish in our experience as human beings, what's the point of hypothesizing about where AI is going to go? I had written about this earlier in the year in a piece named "After the Thaw of Two 'AI Winters,' a Spiritual Spring Heats Up Our 'Consciousness Winter.'" ❦

Then there was the exit of my sixth venture, data.world, in 2025. ServiceNow announced they were acquiring our company on May 7 (7 being my divine number). And then the deal closed on … July 7 (7/7)! Yes, it indeed seems like the divine was telling me I needed to get on with helping humanity have love conquer fear!

I said this book is ambitious, and it really is. It starts with *Definition*: defining what fear and abundance are, then transitions to *Application*: deeply analyzing the biggest challenges and opportunities facing humanity during this exponential of exponentials age, and concludes with *Transformation*: considering the risks of us making it to the Age of Abundance for All, being curious about what our reality actually is, how to take the inner journey to transform yourself as an individual, and then being inspired by models of how humanity has made it through transitional ages to emerge better on the other side and those that are really applying themselves today.

I hope you love this book. I've put a lot into it, and I have a deep love for *all* of humanity. It's up to us, collectively as Team Humanity (with the recognition that we are 99.9 percent the same DNA-wise across this beautiful Earth), to make it to an age that we can barely contemplate today. The coming Age of Abundance is the metaphorical and mystical merging of Heaven and Earth that has been prophesied in ancient spiritual texts as well as prayed for throughout time, and here we are, and it is in this lifetime. We just have to grasp it and use our free will to manifest it. *Love Conquers Fear* isn't just a book, it's the metatheme of how humanity transcends. It's my manifesto. The message of what we have to do is simple to state, but hard in practice.

So, let's dive into Part One of this book on "Definition" and start with discussing that age-old serpent of fear. It's time to slay the dragon.

DEFINITION

Ultimately, there are only two things:
love and fear. Perfect love casts out fear.

—Anthony de Mello, *Awareness*

In the emerging Age of Abundance—already cresting the horizon—our most urgent task will be to finally conquer humanity's oldest companion: fear. Our agency is binary: We choose fear or we choose love. This choice will determine whether the abundance ahead becomes a new liberation or a new tyranny, whether AI and technology serve all people or only the powerful few, whether we retreat behind walls or step confidently into shared possibility. Recognizing this choice is not merely a poetic sentiment; it is the strategic necessity of our time. Every great binary eventually yields to unity—matter and energy, wave and particle, self and other. Both ancient wisdom and modern science point to this truth. This divide will yield, too. But it will require a profound conscious choice as we, all 8.2 billion of us, navigate a dramatic and turbulent bend in the river of our planetary sojourn. We're approaching our own Horseshoe Bend, magnified to cosmic proportions, where the direction we choose to steer will define the next thousand years of human history. The water is gathering force. The bend is here. The current is swift. Snags, submerged hazards, and eddies mask countless vortices in this river. The choice is ours, and it is stark: It's Team Humanity or we perish in the rapids of our own creation.

My good friend and fellow Texan, John Mackey, the co-founder and former CEO of Whole Foods, counsels that courage, in a sense, is *choosing* to overcome fear. "It is fear which prevents most people from reaching their fullest potential in life, fear of failure, fear of rejection from people we care about, fear that we simply aren't good enough, and sometimes even fear of our own potential greatness," John wrote in one of his many essays, and in his autobiography, *The Whole Story.* Ever fearless, John, now in his eighth decade in life, is on yet a new journey to create a network of one-stop, holistic health centers with his new business, Love.Life. This act of choosing is the bridge between the world we have and the abundance we seek.

Broadly, the intent of this book is to explore both the headwaters of this river—those emerging from the thaw following the second consciousness winter, about which I wrote in February 2025, "After the Thaw of Two 'AI Winters,' a Spiritual Spring Heats Up Our 'Consciousness Winter'" —and the new directions this bend in the river is carrying us. We are moving, as both a physical species and as a "Heavenly realm of Light beings," toward a destination where the boundaries between us dissolve.

The elements of this bend are fourfold—the suite of technologies I've dubbed the *Superfecta*; the powerhouse of Artificial Intelligence (AI), quantum computing, and robotics—paired with a more challenging fourth: brain-computer interfaces (BCI). They are the engines of our transformation toward the Age of Abundance for All, but they are powerless without our conscious intent.

This is not, however, the first journey of humanity through a transformative bend in the river of consciousness. History testifies to the fact that what seems permanent is often just waiting for a shift in perspective. The end to slavery, the enfranchisement of women, and marriage equality are just a few of the examples that illustrate how ignorance persists—until it doesn't.

The abolition of slavery was one such transformation. A deviant and disgraceful practice, thousands of years old, it was banned by law and widely shut down across our planet in roughly 180 years. The decisive momentum came between 1780 and 1962. Mauritania was the last country to officially criminalize it in 2007. But the global consciousness had already made its choice: Humanity is not a commodity.

The expansion of voting rights for women followed a similar trajectory: from virtually no countries granting women suffrage in 1870 to over 129 out of 198 countries and territories having done so by 1960. New Zealand led the way in 1893, followed by a cascade of nations, including the United States with the Nineteenth Amendment added to the Constitution in 1920. Switzerland was among the last major democracies to grant women federal voting rights in 1971, followed by full voting rights in regional and local elections as well ... in 1991.

The ongoing nature of this work is exemplified by organizations like The 19th*, a national news organization founded in 2020 that champions women's empowerment.* Its name celebrates the Nineteenth Amendment that gave suffrage to American women, while the asterisk next to its name and logo reminds us that the franchise was not fully extended to many Black Americans until five decades later. As we'll explore in Chapter 8, "The Objectification of Women," this work of empowerment remains unfinished.

Consider same-sex marriage, which seemed like a cultural bridge too far—until suddenly it too wasn't. The Netherlands became the first country to legalize marriage equality in 2001, and today over thirty-eight countries—affecting nations with 1.5 billion people, or 20 percent of the world's population—recognize the same. What was unthinkable became unremarkable in just two decades.

Each of these enormous cultural and political transformations reveals the same pattern: Entrenched systems that seemed permanent for millennia can shift with stunning speed once a critical mass of human consciousness evolves. But to John Mackey's point above, and the work of another friend, the technology ethicist Tristan Harris that I'll introduce soon, this evolution is a matter of our choice—our agency, our free will.

For now, as we reach an even more profound transformation of our consciousness, the bend in the river where science and spirituality align, our choices will determine humanity's destination.

The Discovery of Our Universe's Origin

French Jesuit paleontologist and mystic Pierre Teilhard de Chardin is most famous for coining the term "Noosphere" in his 1922 book, *Cosmogenesis*, for the sphere of human consciousness and the concept of an "Omega Point," the point in time when consciousness and the material world converge as the ultimate point of evolution. In a comment widely attributed to de Chardin, he described this bend in the river a century ago:

> *We are not human beings having a spiritual experience; we are spiritual beings having a human experience.*

The work of Max Planck, the German theoretical physicist who revolutionized our understanding of energy and matter through his foundation of quantum theory, is a subject we'll explore in depth in Chapter 12, "What Is Our Reality?" But it's worth mentioning that this towering scientific intellect and contemporary of de Chardin had similar ideas. Plank famously stated: "I regard consciousness as fundamental. I regard matter as derivative from consciousness."

Understanding this river of consciousness, and navigating the convergence of soul unfolding as we circle this cosmic bend, is a grand journey. So, let's go back upstream to a key tributary to these profound waters: fear.

The Paradox of the Primary Pulse: Navigating Our Ancient Instincts

Fear is fundamental to our very biology: Our "fight-or-flight" response has kept our species alive since we began to walk upright. Fear has driven some of our greatest achievements—from mastering fire 400,000 years ago for safety and warmth, to the seventeenth century scientific revolution born of plague; from landing on the moon in a race against the Soviet Union, to developing an mRNA vaccine in record time when COVID-19 struck.

Fear has profound implications for human evolution. Stanford University archaeologist and historian Ian Morris has proposed a startling paradox: That war—humanity's most destructive impulse—has been a primary driver of our march toward more peaceful societies.

In our Stone Age past, violent death claimed up to 20 percent of all lives. Today, despite our capacity for unprecedented destruction, it accounts for less than 1 percent. In his 2014 book *War! What Is It Good For?* Morris argues that the very horror of conflict—clearly fear—has propelled us forward. It has compelled us to build larger, more stable institutions that make violence increasingly costly and cooperation increasingly profitable. Each terrible war has taught us, inch by bloody inch, that our survival depends on transcending the tribal instincts that once served us.

Neither Morris nor I celebrate war's role in this process, but we cannot ignore a hard truth: Jean-Jacques Rousseau's imagining of inherently peaceful early societies was romantic fiction. The archaeological record tells a different, grimmer story, one that makes our current relative peace not inevitable but miraculous. The question now is whether we will choose to complete this evolution—transcending fear itself—before our weapons outpace our wisdom.

Though we've moved on from the Stone Age to today, fear remains the driving force behind our politics, economy, and even how we think. What once protected us has become our greatest threat. Just as the human body's auto-immune system turning on itself triggers a cascade of pathologies, this powerful emotion—fear—can turn into a deviant feedback loop in the mind that leads to madness, and it leads to rather large scale, distributed madness at that.

It's time to learn how to control this basic instinct before it damns our future. We'll return to Morris, de Chardin, Planck, and other thinkers on humanity's existential turn as we approach this cosmic bend in humanity's development. Along with their ideas, I want to take you forward on a journey to a new, collective consciousness devoid of fear and scarcity, a consciousness of co-creation with the Superfecta of technologies—AI, quantum computing, and robotics, plus BCI. But first, let's go back in history to the headwaters of recorded history and examine this foundational emotion.

The Choice of the Hero: Fear as the "Monomyth"

The *Epic of Gilgamesh*, inscribed in clay more than 4,000 years ago in Mesopotamia, is the oldest known narrative to explore the complexity of fear. While the epic's cosmology is archaic and materialist, its theme is instructive in its familiarity: The hero faces down monsters both literal and existential, and through the act of summoning courage, moves from fear toward wisdom. Nearly 2,000 years later, Homer's *Odyssey* followed this same arc as Odysseus cunningly escaped the Cyclops, resisted the seduction of the Sirens, and withstood the dreamy oblivion of the lotus-eaters. To meet fear with courage, endurance, patience, and strategy is the meta-pattern of our most enduring stories in the millennia since—from *Hamlet* to *The Lord of the Rings* to *Star Wars*, where the hero's true battle against fear is ultimately a contest waged within one's own self.

Fear is the ceaseless foe of heroes throughout history. Mythologist and author Joseph Campbell's concept of "The Hero's Journey" described in his work, *The Hero with a Thousand Faces,* identifies a universal narrative pattern throughout history, found in myths, stories, and religious traditions across cultures. Campbell's "monomyth" throughout his canon of work on myth and symbolism is fear. These stories form the starter yeast baked into virtually every Hollywood blockbuster and iconic children's story. We tell and retell these beautiful tales of the best of humanity, and they are as old as time.

Image 1.1: The oldest known narrative to explore the complexity of fear, *The Epic of Gilgamesh*, as recorded in Sumerian text. (Wikipedia)

To reframe this metaphorically in the language of our day, the ethos of the hero confronting fear is a deep "program" within humanity's "cultural operating system." It is as foundational as Linux, as essential as HTML, and as embedded as TCP/IP itself.

Fear runs deep. It is at the heart of our faith traditions. It is among the oldest spiritual instincts. It is not merely an instinctual reaction to threat but defines our posture before the mysteries of existence. In this way, fear becomes a compass pointing not away from the divine, but toward it.

The Spectrum of Awe: Choosing Our Spiritual Posture

In the great monotheistic religions, fear is rarely just terror; it is reverence—recognition of our smallness before the vast and the unknowable. All the faiths share a similar core of virtues.

In Judaism, the fear of God—*yirat Adonai*—is regarded not as a negative emotion, but as the beginning of moral and spiritual clarity. As the Hebrew Bible declares: "The fear of the Lord is the beginning of wisdom; all who follow his precepts have good understanding" (Ps. 111:10 NIV).

This kind of fear reflects a deep-seated awe, a trembling before the ethical order of the universe, and is a reminder of the humility we should have. For we Jews, fear is not of punishment alone, as we more deeply feel the fear of failing to uphold our responsibilities as creatures in covenant with the Prime Creator. The driving mantra for Judaism—to use that unifying and ancient Sanskrit term for "sacred utterance"—is *Tikkun Olam,* "to repair the world."

In Christianity, this fear is transformed and refracted through Jesus Christ—an invitation to love that still honors the mystery and might of God. The New Testament echoes the Hebrew reverence but moves toward a perfect love that "casts out fear" (1 John 4:18 NKJV). Yet this love doesn't dissolve reverence; it redefines it as a relationship, as grace meeting awe. The driving mantra of Christianity is that the kingdom of God is found within one's self (i.e., in one's soul).

In Islam, *taqwa*—often translated as "God-consciousness" or "piety"—embodies this dual sense of fear and devotion. This fear is not a cowering dread, but a spiritual attentiveness, a state of the heart that remembers God in all things and is wary of stepping off the path of justice and compassion. The *Qur'an* frequently praises those who "fear Allah" as those who act with humility, fairness, and remembrance.

Buddhism offers perhaps the most systematic analysis of fear among all spiritual traditions. Here, fear (*bhaya*) is understood as one of the fundamental forms of *dukkha*—suffering born of our desperate clinging to the impermanent. The Buddha taught that fear arises from attachment—to life, to pleasure, to identity itself—but what we fear losing was never truly ours to possess. Through the *Noble Eightfold Path,* the practitioner learns to observe fear with mindful detachment, neither fleeing from it nor feeding it, but watching it arise and pass away like clouds across an empty sky. In this tradition, fear becomes a teacher—not of divine majesty, but of the illusory nature of the separate self. The ultimate liberation, *nirvana*, is precisely the extinguishing of the fires of attachment that kindle our deepest terrors.

Hinduism, by contrast, approaches fear from an altogether different metaphysical angle. It teaches that fear itself is *maya*—illusion—born of ego and attachment. What we fear losing is not truly ours, and what we believe can be harmed is not the eternal self. The *Bhagavad Gita* 🦅 urges the seeker to cultivate equanimity and detachment, to see beyond birth and death to the *Atman*, the Self or the soul that neither slays nor can be slain. The seeker is advised that true knowledge, *jnana*, dissolves fear like morning light dissolves shadow.

Vedanta, the influential school of Hinduism my wife, Debra, and I studied under the guru Swamiji Parthasarathy 🦅 in Pune, India, takes this dissolution even further, declaring that fear is the product of duality itself, the point I made at the outset of this chapter. In this case, the "binary" is the false perception that there is an "other" to be feared. In the Advaita teachings, all fear stems from *avidya*, ignorance of our true nature. When the seeker realizes through direct experience that the Self (Atman) and the ultimate reality (Brahman) are one, fear becomes impossible—for what could threaten that which is everything?

As Hinduism's foundational text, the *Upanishads*, proclaim: "Where there is duality, there is fear; where there is non-duality, there is no fear." This is not courage in the face of danger, but the recognition that the very framework in which fear operates—the division between Self and other, between threatened and threatening—is itself the fundamental illusion from which we must awaken.

Yes, all the faiths share a similar core of virtues. One ancient faith system, being newly rediscovered in China, Tianxia 🦅, holds promise for our planetary awakening, suggests philosopher Zhao Tingyang—for, as a polytheism, it is already respectful of all the world's gods. In this time of fear, I particularly admire Tianxia's precept, which inverts our Golden Rule to challenge fear of the "other" endemic in our world. Rather than "Do unto others as you wish them to do unto you," Tianxia counsels an ethos more suitable to our age: "Do unto others as they would want you to do unto them."

Choosing Freedom over Coercion

It is, of course, a tragic irony that many native faith traditions, both in North America and elsewhere, were expressed through rites conquering fear, which was precisely the weapon used in many ways to break their communities apart in the centuries of colonialism. As David Graeber and David Wengrow argue in their riveting 2021 book *The Dawn of Everything* , we need a new anthropology that recognizes how many early societies—far from being held together by fear—celebrated freedom, civic debate, and creative play. In fact, the spiritual traditions in these societies often provided frameworks for testing the limits of power precisely to avoid settling into the fear-driven stasis that grips us today.

While some Indigenous traditions do acknowledge spirits who can punish or protect, *The Dawn of Everything* emphasizes that these beliefs functioned within networks of reciprocal obligation rather than unilateral divine coercion. By contrast, colonizers weaponized fear—of famine, epidemic, or spiritual doom—to break community bonds and enforce dependency, including through mission-school indoctrination.

Across traditions, then, we find a common thread: Fear is not always the enemy. At its highest form, fear can be a teacher, a guardian of conscience, and a doorway to humility, reverence, and the search for truth. But, when manipulated—when severed from wisdom and turned inward or outward in distorted form—it becomes something else entirely: a prison of the soul rather than its guide.

A magnificent and transcendent exploration of fear is that of Gary Zukav, author of the contemporary treatise on spirituality, *The Seat of the Soul*, originally published in 1989 in which he writes:

> The human emotional spectrum can be broken down into two basic elements: love and fear. Anger, resentment, and vengeance are expressions of fear, as are guilt, regret, embarrassment, shame, and sorrow. These are lower frequency currents of energy. They produce feelings of depletion, weakness, inability to cope, and

exhaustion. The highest frequency current, the highest energy current, is love. It produces buoyancy, radiance, lightness, and joy.

The Entrepreneurial Leap: Choosing Action over Paralysis

Fear in its many dimensions has certainly accompanied me on my own journey as an entrepreneur. Even though I've been clear since age seven on my divine purpose in this lifetime—to do everything in my power to create a better world through technology—I have had to overcome fear at the beginning of each entrepreneurial venture, seven of them and counting at this point. And at times well after launch, such as the paralyzing fear during the pandemic that I confronted and wrote about in a 2021 series, "A new flu-like virus distant, abstract and then suddenly ... 'This is real, this is bad.'"

As I wrote in my first book, *The Entrepreneur's Essentials*, no matter how experienced you become, fear is always present, especially when you are creating something truly novel. You may fail in a very public way, and the more success you've had, the more that public failure will assault the ego. Unless fear is tamed by the development of higher intellect (which Vedanta teaches), we remain imprisoned by it. The path to entrepreneurial success requires constant motion toward the company's mission. Fear paralyzes action while love animates it—love for the people who have chosen to join your mission, both team members and customers—even in the venture's darkest moments.

Perhaps no one has explored fear as an enslaver of the soul, as a political tool and force, as powerfully as Hannah Arendt, the towering political philosopher known for her chronicle of the 1961 trial in Israel of Nazi Adolf Eichmann that led to her groundbreaking 1963 series in the *New Yorker,* "Eichmann in Jerusalem."

Fear as the Engine of Totalitarian Domination

In her earlier work, foundational for her later insights into the mind behind Hitler's "Final Solution," Arendt argued that fear is both the engine of totalitarian domination and a symptom of modern alienation. In *The Origins of Totalitarianism*, published in 1950 and still resonating geopolitically today, she shows how regimes use fear to isolate individuals, dissolve social bonds, and manipulate reality itself. Fear atomizes people into lonely subjects, more easily controlled, more willing to accept lies, and more easily swept into ideologically sanctioned violence.

In *The Human Condition*, which followed in 1958, Arendt turned to modern life, warning that the erosion of public life and political engagement leaves individuals trapped in private anxiety. Here, fear is no longer just political but existential, fed by technological acceleration and the collapse of shared meaning. Against this fear, Arendt offers a remedy—civic courage—not heroism in the ancient mold, but the quiet bravery to appear in public, speak truthfully, and act in concert with others.

Were Arendt to survey today's landscape—the cognitive addiction engineered by social media, the concentration of algorithmic power in the hands of tech oligarchs, the AI systems that erode judgment while amplifying bias, the data centers treating human attention as extractable resource—she would see the fulfillment of her warnings. So much in the private realm has been weaponized: our anxieties are being harvested, our isolation is monetized, and our capacity for thoughtful deliberation has been replaced by algorithmic reflex. She would probably argue the public square has fragmented into personalized echo chambers where disinformation flows along the paths of least resistance, and shared reality itself becomes negotiable. What she feared most—the collapse of the space where humans could appear to one another as equals and act in concert—has been engineered into the infrastructure of modern life. Her remedy—civic courage—remains.

But civic courage now means something harder: the discipline to resist systems designed to capture us, the bravery to seek out genuine plurality rather than curated sameness, and the commitment to rebuild public spaces where political action—not mere reaction—becomes possible again.

We'll discuss this further in "New Models for Humanity" in Part Three, "Transformation." Model 4 in that section is on the Cosmos Institute, "Where AI's Suite of Technologies Is a Verb and Not a Noun."

But let's get back to Arendt for now. In a new introduction to the latest edition of *The Origins of Totalitarianism,* republished in 2024, the Pulitzer Prize-winning author Anne Applebaum frames fear not only as a historical force in the rise of totalitarianism but as an urgent, contemporary challenge. Applebaum writes: "Reading [Arendt's] account now, it is impossible not to wonder whether the nature of modern work and information, the shift from 'real life' to virtual life and the domination of public debate by algorithms that increase emotion, anger, and division, hasn't created some of the same results." Similarly narrowing in on more recent decades, sociologist Barry Glassner has examined how modern media, in collusion with political and commercial interests, has manufactured a climate of pervasive but misplaced fear. In *The Culture of Fear,* which came out in 1999 toward the end of the public internet's first decade, Glassner exposed how Americans have been led to fear the wrong things—child abductions rather than car accidents, terrorism rather than poverty, violent crime rather than medical error. These fears, endlessly magnified by a 24/7 media cycle—and now led by algorithmically engineered social media—hijack the public imagination and distort our sense of risk. The result is not merely anxiety but paralysis, a kind of civic disorientation that echoes Arendt's concern: Fear as the enemy of democratic action and rational judgment.

More recently, Harvard psychologist Steven Pinker has taken Glassner's diagnosis of misplaced fear a step further by showing that the world, by almost every measurable indicator, has become significantly less violent, yet our perception of danger has never been more acute. In *The Better Angels of Our Nature* in 2011 and *Enlightenment Now* in 2018, Pinker marshaled an avalanche of data to argue that the rates of homicide, war, famine, child mortality, and even domestic abuse have declined dramatically over the centuries and continue to fall. Yet surveys consistently show that people believe the world is getting more dangerous, not less. Why this disconnect between reality and our perception? Pinker points to a combination of media incentives, psychological biases, and political manipulation. Bad news is more salient, more urgent, and more marketable than good news and our brains, still wired to detect threats, are far more attuned to stories of disaster than slow, statistical progress. This means our fears are not simply exaggerated, they are systematically skewed. We overestimate the risk of airplane crashes and terrorism but underestimate the ongoing toll of climate change or antibiotic resistance. Pinker sees this as not just a psychological error, but a moral and civic failure. When we fixate on vivid, anomalous threats, we neglect the quieter, chronic dangers that require sustained collective action. His work reinforces both Arendt's concern about the erosion of public rationality and Glassner's critique of fear-driven narratives. But Pinker grounds the problem in a broader Enlightenment ethos, suggesting that human progress is real, but fragile, and that our perception of fear must be recalibrated to match reality if we hope to sustain that progress. We'll return to this Enlightenment ethos in the next chapter, "Abundance."

Recalibrate we must. But fear, and the ways in which we respond to it, is not merely a cultural phenomenon. It is biological. At the center of our fear response is the amygdala, a small almond-shaped cluster deep in the brain that evolved to trigger instant reactions to danger. This was useful when predators lurked in the brush, but in a world saturated by media, symbols, and abstractions, the amygdala's snap judgments often misfire. It cannot tell the difference between a lion in the tall grass and a political advertisement, a distant news story, or a TikTok panic spiral. Critically, the chemically-driven response is instantaneous upon perception of a threat. But once the amygdala's dose of fear is pumped into our senses, the deep fear lingers, like Red Bull keeping you awake long past the exam that you amped up for that is now thankfully over. Our most ancient brain structures, now infused with what I call "synthetic" fear have not caught up to our technological landscape. For all of the good of global connectedness, thanks to the internet and the Web, we are now dealing with the most intelligent (and manipulative) algorithms ever produced to hijack our amygdala at scale. To Applebaum's point in her review of Arendt, AI can accelerate this hijacking even faster, or it can be used to *combat* hijacking our brains even faster—if we make that choice.

As evolutionary biologist Edward O. Wilson noted in a famous talk in 2009: "The real problem of humanity is the following: we have Paleolithic emotions, medieval institutions, and god-like technology... And it is terrifically dangerous, and it is now approaching a point of crisis overall."

Image 1.2: Edvard Munch's "The Scream," reimagined with AI and yielding an apparent third hand.

The technologies that surround us—from WiFi to GPS to social media—are marvels that amplify our connectivity, but they also amp up our collective dread. Economic displacement affects virtually every country, and 122 million refugees have no country at all. From wars, to natural disasters to horrific crime, we feel the shock of events across the globe instantly and profoundly. This hyperconnectivity creates a Möbius strip of anger that circles the world, amplifying feelings of instability, fear, and dread, feeding into what I call the scarcity mindset—the constant anxiety that anticipates imminent loss.

Fear as a Form of Capital

This crisis of mismatched perception sits precisely at the fault line of this imbalance between fear and our biologically, evolutionarily, technologically, and now algorithmically fueled perceptions of fear's drivers. Fear is an ancient survival mechanism, manipulated by institutions out of step with modern realities, and amplified by technologies we have yet to master. Fear, which is so primal and yet so malleable, has become a central force in our politics, media, minds, and myths. In fact, fear has become a form of capital, a critical input to our data-driven economy.

The Web's advertising-driven business model has gradually fine-tuned algorithms to maximize engagement through enragement. Though founders like Sergey Brin, Larry Page, and Mark Zuckerberg once resisted this approach, they—and many other tech titans—now profit handsomely from systems that exploit our most primitive instincts.

We now find ourselves at a crucial juncture where fear is not merely a byproduct of media sensationalism or political propaganda—it is actively engineered and harvested at scale. This is profoundly evident in the work and warnings of Tristan Harris, speaking out on the dangers of AI at TED's 2025 gathering. I've known Harris for some time and am a fan of the 2020 documentary *The Social Dilemma* in which he features so prominently. The key takeaway from the movie is that social media is both a utopia and dystopia and its machine-learning-driven network effects are nearly impossible to overcome due to the hijacking of our offline social networks at online scale.

The "Choice Point" of AI

Harris, a former Google design ethicist and co-founder of the Center for Humane Technology, has become one of the most important voices deconstructing how technology companies—particularly social media platforms—exploit the human fear response for profit. Ultimately, this is a choice, Harris argued, one made through our neglect of many warnings on the negative impacts of social media: addiction, distraction, polarization, and harm to mental health, particularly among youth. Harris argued that the world now faces the same kind of critical "choice point" with AI.

While the *possible* with AI is nearly unlimited potential—solving major problems, accelerating progress across every field—Harris has stressed the importance of discussing the *probable* and asking: What is likely to actually happen given current incentives? He cautioned that the relentless competition to achieve dominance in AI development is incentivizing companies to take safety shortcuts and release increasingly powerful but insufficiently controlled systems.

His central argument is sobering: Today's attention economy doesn't just commodify our time or data, it commodifies our most primal impulses. The systems we have built do not merely respond to what we click on; they are architected to provoke, sustain, and amplify emotional arousal, especially of the fearful kind. It should hardly be surprising that the Oxford University Press chose "rage bait"—defined as "online content deliberately designed to elicit anger or outrage by being frustrating, provocative, or offensive"— as its 2025 Word of the Year.

Harris came to prominence with his metaphor of "the race to the bottom of the brainstem," a vivid phrase capturing how companies optimize their algorithms not to serve human flourishing but to once again hijack our amygdala. What the ancient brain once used to spot predators, the modern internet now targets with tailored outrage, tribal content, and if-it-bleeds-it-leads logic encoded in the algorithms. The result is a technological feedback loop in which fear becomes the primary fuel of engagement—and engagement becomes the benchmark of success.

A major culprit is a 1996 law known as Section 230 that exempted then-nascent internet companies like CompuServe and America Online from legal liability for content carried on their networks. That was understandable at the time—analogous to the phone company not being liable for plots or threats made over its network—but this was *before* algorithms began amplifying the pernicious concept of "engagement."

"Any single video promoting extreme dieting might not pose serious risk to a viewer," wrote entrepreneur, academic, and technology critic Scott Galloway in one of his many critiques of Section 230, "but when YouTube draws a teenage girl into a never-ending spiral of ever more extreme dieting videos, the result is a loss of autonomy and an increased risk of self-harm. The personalized feed, churning day and night, linked and echoing, is a new thing altogether, a new threat beyond anything we've witnessed."

Galloway has proposed changing the law. It could still protect neutrally transmitted content, but not content that has been algorithmically generated or amplified.

As Harris has often put it: "If you're not paying for the product, you are the product. And the product is your emotional vulnerability." His work connects squarely with earlier thinkers such as Arendt and Glassner. Where Arendt warned of fear's use by authoritarian regimes to isolate and control, Harris reveals how fear is now wielded by invisible, decentralized systems—algorithms we neither see nor understand, yet which shape our perceptions, habits, and even our political identities. Where Glassner exposed the media's penchant for exaggerated risks, Harris shows how this logic is now encoded in the very infrastructure of the internet, where every click feeds back into a system that learns, iterates, and adapts—not toward truth or wisdom, but toward whatever elicits the strongest emotional hijacking, the most engagement.

Harris has repeatedly emphasized that we are not prepared for the asymmetry of this contest. Billions of dollars, a global "datasphere" of 175 zetabytes, and some of the best minds in Silicon Valley, are deployed and devoted to studying and shaping our behavior, while we check our phones on the way to the grocery store or in line at the airport and think we're in control. The truth, Harris warns, is that we are outmatched. Our minds, still tuned to the evolutionary settings of scarcity and threat, are no match for machine-learning models trained on petabytes of our emotional history.

In this light, fear has become not just a public health concern or a civic toxin, it has become a resource to be mined, a currency that drives valuation and venture capital, a lubricant for the gears of a new kind of extractive industry. We have entered a *synthetic fear economy,* in which fear is manufactured, optimized, and monetized at scale. This is a profound shift. Just as fossil capitalism drew its energy from ancient carbon, today's platforms draw theirs from ancient biology.

And just as the carbon economy has reached its ecological limits, so too the fear economy may be reaching its cognitive and spiritual limits. Here, Harris's work turns from critique to vision. He advocates for a humane technology movement, one grounded in ethics, designed for human dignity, and aligned with long-term societal well-being rather than short-term metrics of growth or engagement. In short, he calls for what might be thought of as a new moral architecture for the digital age. Harris's call aligns powerfully with a broader, deeper trend—an emerging recognition that we are not merely facing a technological revolution, but a consciousness crisis. The tools we build are mirrors of ourselves, and if those tools are addictive, divisive, and fear-inducing, it is not simply a failure of engineering, it is a failure of intention, of wisdom, and ... of soul.

This is the bend in the river of our evolution as a species.

**Image 1.3: Horseshoe Bend on the Colorado River
is what scientists call an "entrenched meander,"
the result of geological uplift that mirrors
our ongoing Superfecta acceleration.
(National Park Service)**

As we move through the bend, the endeavor to manifest an Age of Abundance for All catalyzed by AI, quantum computing, robotics, and BCI, or the Superfecta, we are not just building machines—we are revealing ourselves.

Let's further detail the elements of the Superfecta.

We tend to discuss and think about AI as a singular notion, but artificial intelligence is an ecosystem of technologies as diverse as the flora and fauna of the Great Barrier Reef. Think of AI in terms of the Standard Model of particle physics, where the nucleus of each atom is made up of different numbers of neutrons and protons that define the character and nature of the elements they form. AI is made from these other three components—quantum computing, robotics, and BCI—and these foundational components combine onto an infinite variety of forms that determine the essential nature of the AI they produce.

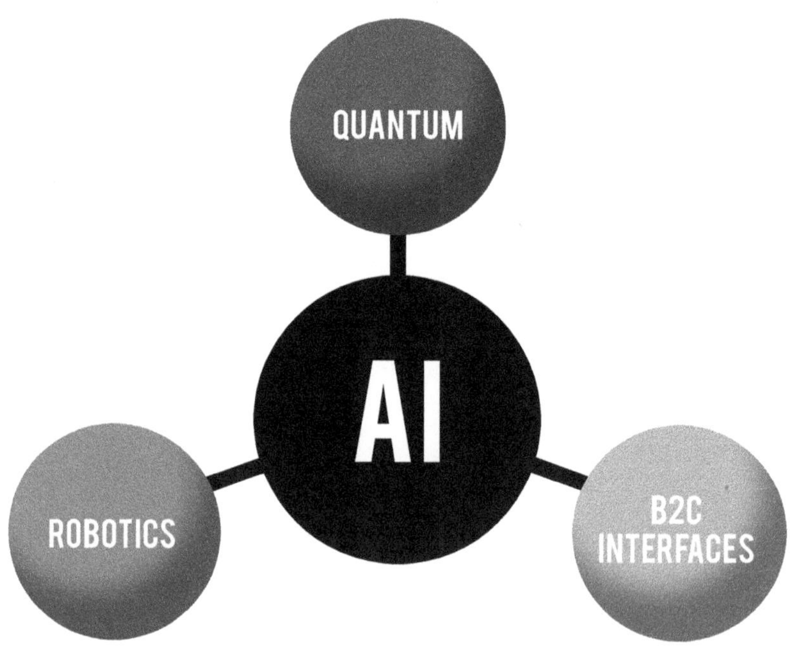

Image 1.4: The Superfecta of a plural lattice of interacting technologies.

And stay mindful that, as Chris Wiggins and Matthew L. Jones put it in their marvelous book, *How Data Happened*: "Some fields, like biology, are named after the object of study; others like calculus are named after a methodology. Artificial intelligence and machine learning, however, are named after an aspiration."

A Poor Compass for Creation

As we scramble to find our posture toward this latticework of technologies, the irony is that it is *not* the technology we should fear. Rather, the greatest threat we face is human unconsciousness that has failed to keep pace with its own power. What we need, in the face of this synthetic fear economy, is not more data but deeper insight. We do not need more engagement but more reflection. And, more than anything, we must summon the courage to imagine a world that no longer runs on fear but instead one where love finally conquers fear.

We need to collectively manifest a beautiful world where love finally conquers the scarcity mindset that has dominated our psyche for millennia. We need to catalyze our world so it becomes one where Heaven and Earth merge, as has been prophesied since the beginning of religion.

The path to our collective, divine calling, exponentially accelerating with the Superfecta, is a challenging one. Today, that path winds through the fiercest and fastest arms race in human history, not a race for land or arms, but for minds, jobs, and the "prize" of global domination. A manic global sprint is underway between the United States and China as both seek to claim dominion over the other, through mastery of artificial general intelligence (AGI), a technology that will first replicate and then exceed human intelligence, leading to what is now being called artificial super intelligence, or ASI. At stake, we are being told, is not just national security or economic primacy, but the very future of civilization.

The U.S. Stargate Initiative, announced on January 21, 2025 by President Trump, is a $500 billion endeavor whose name echoes science fiction. China's DeepSeek R1 model, announced just one day earlier on January 20, is a leaner but symbolically potent challenger. The development of both models is driven by the same underlying fears: fear of falling behind in the global economy, fear of obsolescence as a country, and fear, above all, of being the first to blink in a scarcity-fueled, zero-sum contest for technological supremacy.

But fear is a poor compass for creation and a deal breaker for global collaboration. With the realization of AGI and then ASI, global collaboration, instead of competition, will not be optional. It will be imperative. Humanity has never collectively created a technology this powerful and soon to be as ubiquitous. AGI and then ASI will be nearly free in our near future. ASI, "smarter" than all human minds combined, will be in the hands of 90 percent of humanity, from our best actors to our worst.

The ethos we need is suggested by JD Messinger, who wrote the Foreword to this book, in his own moving work, *11 Days in May*. A nuclear scientist and former officer aboard a nuclear submarine, JD has spent a lifetime grappling with the moral weight of powerful technologies. His conclusion is deceptively simple: Ideas themselves are not evil. Even the most dangerous technologies are, at their origin, morally neutral. What matters is what humans choose to do with them—and the intentions that animate their use, which shift with time and perspective. Nuclear weapons were not born evil; they became so through fear, rivalry, and the stories nations told themselves about survival. "All ideas are good," JD writes, "but what man does with this knowledge and his intentions for doing it may be non-good."

Artificial intelligence now confronts us with the same mirror. The question is not whether the technology itself is good or bad, but whether humanity can mature fast enough to wield it—this time without allowing fear to write the script once again. The challenge for us to collectively manifest our better angels has never been more important. Our spiritual suffering must be overcome and our leadership cores rebuilt, as I'll discuss in the chapters of Part Three of this book, "Transformation."

This is not the first time we've sprinted toward a frontier we barely understand. Twice in the history of artificial intelligence—first in the 1970s and again in the 1990s—hype outran capability, investment dried up, and a so-called AI winter set in. But as the Stargate Initiative and DeepSeek's open-source-fueled rise make clear, we are now well past any third such winter. The snow has melted and the race is on. Yet, while technological innovation accelerates, a colder, more dangerous freeze may still grip us: a third "consciousness winter," a term I coined in early 2025 to describe past stumbles in our striving to overcome biologist Wilson's three-part dilemma.

The Consciousness Spring

I used the term *consciousness winter* to mark our intellectual and cultural retreat from the deeper, older, and more urgent question: What is consciousness? And can we create machines that think, feel, or know, when we ourselves scarcely understand what consciousness is? In the mid-twentieth century, as behaviorism and materialist science gained sway, introspective studies of consciousness were cast out of mainstream discourse.

The pioneers of the study of consciousness, of course, were William James in the late nineteenth and early twentieth centuries and Aldous Huxley, a generation later. Both believed our individual consciousness is the filtering down of a larger universal consciousness, conclusions they both arrived at through research and scholarship into the ideas of cosmogenesis formed by de Chardin, but also in part through their use of psychedelics. Their work, however, was eclipsed by the triumph of cognition over contemplation as the right posture of science. Though today, as neuroscience and quantum physics circle back to questions of consciousness, their insights feel less like mysticism and more like prophecy.

A perfect storm of fear-based thinking, politics, and policy making, from President Nixon's War on Drugs to the battles against the cultural revolutions of the 1960s waged by President Reagan and Britain's Margaret Thatcher, drove psychedelic research and medicine, along with much exploration of spirituality, from universities, the media, and polite discourse. The capstone of this enshrouding was really Daniel Dennett's *Consciousness Explained* in 1991. It did not explain consciousness so much as reduce it, attempting to explain it away entirely.

But just as AI has emerged from its own winter, a consciousness spring now stirs. A series of events in the mid-aughts marked the moment "when sensitive ears could make out the sound of the ice beginning to crack," wrote author Michael Pollan in his seminal 2018 book, *How to Change Your Mind.* Psychedelics were returned to the focus of scientists at universities and research centers around the world, including at Johns Hopkins University in Baltimore and the University of Texas at Austin, in the pioneering Dell Medical School. Spiritual inquiry and scientific investigation, long estranged, are finding each other again. New thinkers—author Mark Gober, near-death-experiencer Eben Alexander, and such people as Pollan—are reviving a truth that ancient traditions never abandoned: Consciousness is not a mere byproduct of neurons firing, but a primary force in the cosmos.

This is what philosopher James Ogilvy calls a "third way" in his new book, published in 2025, *It's All Coming Together—Essays on Emergence*, in which he argues that neither secular reductionism nor dogmatic belief, but something more mysterious, and more unifying, is pulling us together.

Pollan's book, which includes his own accounts of journeys enabled by a new kind of medicine, is a masterful survey of both the rejection of psychedelics by science, and now its rediscovery, not just for treatment of depression, PTSD, and other tough maladies, but also as a window into the true nature of consciousness. Pollan writes: "A handful of scientists working in both Europe and the United States are opening a new window into consciousness, and what they are glimpsing through it promises to change our understanding of the links between our brains and our minds."

This revival of scientific interest in psychedelics comes not in opposition to AI, but as its unintended consequence, for as we try to replicate consciousness in machines, we are forced to ask what, exactly, it is we're trying to replicate and reckon with that which makes us most human. In doing so, the boundaries between science and spirituality begin to blur. Quantum entanglement and near-death experiences, neural networks, and mystical insight all begin to point, in strange harmony, toward something larger than ourselves, something we do not yet understand, but feel drawn to.

I want to be clear upfront on what I believe: Consciousness is fundamental and is the very nature of the universe itself, with love encoded as its highest value.

Humanity's Most Fateful Choice

So, where does that leave us? Humanity stands at a crossroads, not just of technology, but of meaning, and we are also at a crossroads of choice.

I believe I know what Arendt might counsel—particularly through Applebaum's refined update of her views—for the vicissitudes of today's suite of AI technologies. If we continue down the path of manic fear, we risk creating tools that far outstrip our moral readiness to wield them. Arendt and other wise thinkers—from Harris to Glassner to Wilson—offer a stark warning: The totalitarian impulse doesn't require a single dictator anymore. It can be distributed through systems—through algorithms—that isolate us into personalized feeds, through AI that amplifies our worst instincts while claiming neutrality, through platforms that monetize our atomization. What Arendt feared most—the collapse of the space where humans could appear to one another as equals and act in concert—is now being engineered at scale. The public square has been replaced by privatized digital infrastructures where we are not citizens but data points, not participants but products.

But Arendt also gave us the remedy, civic courage and Applebaum makes this explicit for our moment: Civic courage is not the heroism of mythology, but the daily practice of choosing to act with others rather than being acted upon by systems. The choice before us is whether we will allow AI to become another instrument of isolation and control, or whether we will insist that these technologies serve human dignity, genuine plurality, and the possibility of collective action.

This moment of decision is the essence of our Horseshoe Bend. We stand not merely at a technological threshold but at a moral and existential one. JD Messinger makes the same point: With the creation of any technology, the application and intentionality of its use deeply matters. We can build intelligence not only in our machines but in ourselves. We can ask deeper questions, reintegrating wisdom traditions with scientific exploration. We can chart a course not of domination and subjugation, but of discovery and liberation; not one of fear, but of love; not of separation, but of unity and oneness.

The choice is ours, but the window for making it consciously is narrow. As the Superfecta exponentially accelerates faster and faster, we must decide: Will we allow fear to engineer our future, or will we summon the courage to act in concert, to rebuild the public spaces—both physical and digital—where genuine human plurality can flourish? This is not command, but comprehension; not conquest, but co-creation. The challenge will be to wield the enormous power of AI with care and build a future of Abundance for All that transcends the scarcity mindset that has imprisoned us for millennia.

> *Love is the opposite of what an atheist believes, and that is God. God is love. All energy is constantly seeking a higher state of evolution, the purpose of change. To artificially sustain a subject of purpose would be to shut down the change process and the very circle of life.*
>
> **— JD Messinger, *11 Days in May***

On the way to the talk I gave to the fifty CEOs at MIT that I mentioned in the Introduction, I was met at our home for the ride to the airport by an Uber driver at around 4:30 AM. The cheerful driver's name is Kallibou. He's an immigrant from Chad, a Saharan nation three times the size of California.

Kallibou and I immediately connected and started talking about Chad and his life journey. As we did so, I used ChatGPT to learn about his country, which numbers 187 out of 189 countries on the Human Development Index (HDI). Chad is a country with a GDP of around $660 per person. Compare this with the United States at almost $86,000 per person.

We had a moving conversation about how his wife and children, who are still in Chad, are faring. We spoke of how much he is financially supporting them and his parents through his job in America. He spoke of America, and being in America, with deep pride. He shared how much he wants this life for his children and wife and how they would hopefully be moving here after six years of separation as he saves for that day.

It struck me that a country like Chad will only be helped by the accelerating forces of AI—quantum computing, and robotics, plus the brain-computer interfaces (BCI)—those exponential forces forming the Superfecta we discussed in Chapter 1, "Conquering Fear." I imagined a non-governmental organization (NGO) of the future enabled by AI to build all of the clean water plants and pipelines that Chad would need. That NGO, or a startup business, would be able to help the people of Chad develop the land for farming and nearly unlimited food of the highest quality. It would harness the power of the Sun—a topic we'll explore further in Chapter 4, "Climate Change"—to provide virtually unlimited energy, which would flow throughout the country, transforming it by delivering electricity to every home, even in the most remote of areas. The NGO would be able to provide nearly unlimited healthcare and the cure for most diseases, the treatment of which until now would have been confined to the most developed regions of the world. Chad, which is now referred to by pessimists as the "Dead Heart of Africa," would become an oasis. And so would other deeply struggling countries on this sprawling continent of Africa, like South Sudan, Burundi, the Central African Republic, Somalia, Niger, Malawi, Mozambique, and Liberia.

As much as I love America, whose Statue of Liberty shines her lamp to the world, Kallibou's family wouldn't need to move here if these new options were to succeed in transforming their homeland. And, as much as I love the inviting spirit of the line from Emma Lazarus' poem, *The New Colossus*, inscribed at Lady Liberty's base: "Give me your tired, your poor, your huddled masses yearning to breathe free, . . ." I can't help but imagine a world with no tired, no poor, no huddled masses at all. We can build that as Team Humanity.

Perhaps this bright and soulful human being working as an

Uber-driver would be part of the leadership of that NGO. Perhaps he will help transform the poorest countries in Africa to be oases for themselves and the world.

A New Mindset

Humanity's liberation from our longtime companion of fear, driven by a scarcity mindset, which in turn is a product of millennia of human competition for limited resources, is what I mean by the Age of Abundance for All. That reality of scarcity will soon be in the past, but will our mindset be in the past too? We have a social and psychological commitment to scarcity, so abundance alone is not enough.

I often think of futurist Alvin Toffler's influential book *Future Shock,* which I read in college to help my wife, Debra, in a class. It was named and published in 1970. In it, he warned of the disorientation and stress we would experience with the impending struggle to adapt to too much change happening too quickly. He got a lot right, like his prediction of the exponential rise of media and communication technologies that would flood society with data and opinions, leaving people overstimulated and unable to make sense of the world. Now, the future is coming at us at ten times the velocity that even Toffler imagined. Today, we are presented with a future that defies any accurate definition, but which can become an epoch of boundless opportunity, if we understand and embrace an alternative abundance mindset. But, if this is to happen, it must be the Age of Abundance *for All,* and I italicize *for All* to stress this point. Yes, when I talk about an Age of Abundance for All I mean countries like Chad too.

As Kallibou spoke, the contrast between his current reality and the potential of his homeland became clear. For Kallibou, the "Age of Abundance" isn't a tech-utopiast buzzword; it is a practical lifeline that costs about $20 a month. I encouraged him to subscribe to tools like ChatGPT or Anthropic's Claude immediately. These aren't just apps; for a man from a nation with a $660 GDP, they represent a democratization of intelligence that can help him and others in three critical areas of development:

- **Universalize Education:** In this coming age Kallibou will be able to provide his children tutoring in any subject in their own language, as well as instruction for other languages they might wish to learn, and all these lessons will be comparable to those offered to the students at the best universities in the world.
- **Unlock Capital**: Kallibou will be able to access capital to start a business wherever he might choose as AI-powered micro-lending platforms make lending available to entrepreneurs and businesses who would have been shunned by conventional banks.
- **Empower the Community:** Kallibou will channel his drive into leadership, perhaps to lead that decentralized NGO or business that will solve Chad's water and energy crises with machine learning and robotics.

By grounding the abstract theories of human progress in Kallibou's journey, we see that the compounding of knowledge—described by eighteenth century economist Anne Robert Jacques Turgot, which we'll explore in more detail below—has finally reached a velocity where a rideshare driver can wield the same creative power as a twentieth century nation-state.

Kallibou's story suggests that the Age of Abundance isn't about where he lives but the tools he carries. While the Statue of Liberty remains a beacon of hope, the goal of this new era is to ensure that a flourishing life is no longer a matter of geography. By democratizing access to knowledge and capital, we can ensure that the "oasis" isn't just a destination to reach, but an environment that can be built anywhere on Earth.

This liberated age is within humanity's grasp if we allow love to conquer fear. And that happens by changing ourselves from the inside, by allowing our souls to become more embodied in our physical bodies. I'll be discussing that in depth in this book's Part Three, "Transformation."

Building on a Historical Foundation

Broadly, the foundation of the Age of Abundance for All has been long in the making. It results from the compounding of human knowledge driving human progress. This foundation of knowledge is growing exponentially, and just what we will do with this knowledge is being passionately debated. But the insight of Pierre Teilhard de Chardin, the Jesuit priest and scientist you met in the last chapter, is instructive no matter what faith or no-faith tradition you follow. He wrote of an "Omega Point," the convergence of all consciousness into a unified, transcendent state, which I believe is near.

Against the backdrop of this metamorphosis, Kallibou's story isn't just personal—it's part of a much larger historical pattern that philosophers have been tracking for centuries. The modern abundance movement is the Enlightenment's faith in progress updated for the age of exponential technology. Two and a half centuries ago, the economist and philosopher Anne Robert Jacques Turgot was the first to argue that human knowledge compounds like interest, each generation building on the last to turn scarcity into sufficiency. While little remembered today, Turgot's 1750 book, *A Philosophical Review of the Successive Advances of the Human Mind,* formed the shoulders on which stood many of the famous Enlightenment philosophers, including Adam Smith and Thomas Jefferson.

Let's fast-forward from the eighteenth century to today's world through a familiar series of power amplifiers: the productive work of society then was accomplished by 1x of human muscle, this output translated to 10x that with animals, then 100x with steam, then 1000x with solar, then 10,000x with nuclear, and potentially is now unlimited with the advent of fusion, which is just around the corner, a subject we'll get into in Chapter 4, "Climate Change."

Today, nearly three centuries later, many are making Turgot's argument. Most importantly among these is Peter Diamandis, author of the 2012 book *Abundance—The Future is Better Than You Think.* I wrote the longest book review of my life on my *Lucky7.io* blog, 🦅 thanks to that book. More than a decade later, many have embraced the term "Abundance" and its promise.

Most recently, author Ezra Klein has reshaped the idea in his book, which is also named *Abundance*. In it he frames the case as one essentially against red tape. In Klein's rendering, abundance is a practical, civic challenge of building enough—housing, energy, infrastructure, and institutions—to meet human needs at scale. Diamandis himself is often reminding us of this in his brilliant *Moonshots* podcast. Between the time of Turgot and today's movement led by Diamandis, the abundance curve has grown much steeper, the network more global, and the tools are nearly powerful enough to make sufficiency into surplus at exponential scale.

As Diamandis' co-author, Steven Kotler, framed it on the *Moonshots* podcast: "We have to learn to cooperate at scale or else we die." Or, in my nomenclature, love has to conquer fear. In short, the abundance movement is the Enlightenment's progress doctrine in its twenty-first-century form, grounded in Turgot's insight that human reason and exchange—given time and freedom—transform scarcity into sufficiency, and sufficiency into surplus, or... Abundance.

Back to today's reality, there are many signs that the Age of Abundance for All really is just around the corner, not only for the Elon Musks and Sam Altmans, but for Kallibou and his family too. The family of technologies that are now able to assist Kallibou for the cost of a Netflix subscription are poised to be so much more transformative of his reality. Yes, I was encouraging Kallibou to subscribe to ChatGPT or Anthropic's Claude as soon as possible. Together those tools will democratize access to education for his children, enhance healthcare and financial services for his family, and open entrepreneurial opportunities for him on a scale he's never before imagined. Perhaps it won't take as long for Kallibou to save up enough to bring his family to America. Perhaps he could start a business off his phone while waiting for fares at the airport. Perhaps Kallibou could even start a big business and be one of the wealthiest among us in this coming Age of Abundance for All.

Yes, a transformation unlike any before is on the horizon: The power of AI will not concentrate wealth just among the few, it will distribute abundance among the many. With the power of AI in hand, Kallibou's pathway to prosperity could transcend traditional barriers of geography, language, and circumstance. His perspective is one of being a man from Chad. Like so many immigrants, he already sees abundance all around him in America. He's so proud to be here, and I'm sure the journey to make it here wasn't an easy one. With the cost $2,400 to $4,000 to fly roundtrip from Chad to where I live in Austin, his plane ticket alone pencils out to between four and eight years of the average Chadian's income. I know what he'll choose—and it *is* a choice, for all of us.

However, as we discussed in Chapter 1, "Conquering Fear", this intentional choice for a future of abundance, and our readiness to create that future, are framed by the fear of AI and robotics that is deep in our collective psyche (think of the films *The Matrix* or *The Terminator*). Even among some very smart people, even among some friends and colleagues in the AI space, there's a view that the only winners in AI will be the Musks, the Altmans, and the other tech titans of the world. They suggest we're on the cusp of a new kind of hegemony, a techno-feudal order dominated by the Lords of AI. You've seen the movies, and I'm sure at some point you've *felt* the same fear. But that fearful view is certainly not shared by Kallibou, nor by many others who are facing real scarcity or have thought more deeply about overcoming it. In short, many immigrant experiences are like Kallibou's, not sharing in the "synthetic fear" we discussed in the last chapter.

Africa Leads the Way with AI

Without diminishing the real and difficult challenges that will come with this metamorphosis, a subject we'll cover in depth, let's better understand what is already happening on Kallibou's ancestral continent as we approach the Age of Abundance for All.

Scientists in Africa are already leading the world in development of "small language models," or SLMs, that meet local needs. One standout example is InkubaLM, developed by Lelapa AI. It's Africa's first multilingual SLM, designed to support African languages like isiZulu, Yoruba, Hausa, Swahili, and isiXhosa. Soon InkubaLM will support many of the 100 languages spoken in Chad, including my new acquaintance's native Ngambay. Using a technique known as quantization, these Africa-specific models effectively compress the computational power of AI to allow use on entry-level smartphones, even without constant internet connectivity. These lighter models enable practical applications—education, climate info for farmers, multilingual chatbots, and telemedicine.

But, as the innovative technology website founded by Sophie Schmidt, *Rest of World*, has argued through its deep reporting on stories not often found in the mainstream technology press, technology alone cannot make a radical difference. When kids in Colombia began using Meta AI in their phones, they started failing exams. In 2025, South Korea rolled back a plan to introduce AI-powered school textbooks because of inaccuracies and other issues. Social innovation paired with local knowledge and insight is critical as well, such as the public-private partnerships pioneered by the pan-Africa NGO Elimu-Soko, which embeds technology solutions within existing government systems and culturally grounded teaching practices. With assistance from the Denmark-based Hempel Foundation, Elimu-Soko, meaning "education market" in Swahili, launched in Rwanda and Zanzibar in 2021. It is now expanding to Ghana, Nigeria, Kenya, Tanzania, Ethiopia, and South Africa, as detailed by the *Stanford Social Innovation Review*.

**Image 2.1: Girls dive into AI and robotics
at a tech bootcamp in Rwanda.
(Courtesy of the United Nations)**

AI is also transforming financial access. In Kenya, AI-powered platforms like Tala 🐦 and Branch International 🐦 use machine learning to analyze smartphone data—call patterns, app usage, even social connections—to provide instant microloans to people without traditional credit histories. These algorithms can assess creditworthiness in minutes rather than months, offering small business loans that help immigrants and others start ventures impossible to finance through traditional banks. Similar platforms are expanding across Africa and to immigrant communities worldwide, creating economic opportunities that would have been unimaginable just a few years ago.

Moving out from Africa, readers here in America will be well versed in automatized factories, chatbots for health care, or may well enjoy self-driving cars. Beyond the innovations before our eyes, in early 2026 Google Earth Engine (GEE), in collaboration with Google DeepMind, has introduced AlphaEarth Foundations, a groundbreaking foundation model. As described by the developers, the system generates what it terms an "embedding"—a compact, multidimensional summary that integrates years of global Earth observation data from optical and radar satellite imagery to climate simulations, LiDAR, and more—into a unified digital representation. Each embedding captures the characteristics of a 10 × 10 meter patch of land, producing a distinct "digital fingerprint" that is efficiently stored and continuously updated through the Satellite Embedding dataset available in GEE.

Environmentalists in Louisiana struggling to save the Mississippi Delta's wetlands 🐦, which are now disappearing at the rate of a football field every 100 minutes, can instantly scan and compare the state of wetlands five years ago with that of today. With a click, they can instantly identify and examine every similar wetland on the planet and map their dynamics. In Kallibou's homeland of Chad, this same technology could help reverse the shrinking of Lake Chad, which supports agriculture and fisheries for 30 million people—potentially transforming the economic conditions that drove him to emigrate to America in the first place. The implications for scientific collaboration, particularly to confront climate change, are beyond astounding.

I didn't ask Kallibou about his immigration status, but I'm sure he faces bureaucratic challenges—or worse. At Stanford, teams working at the intersection of law, design, and technology have been developing AI-based legal-assistance tools—chatbots and large language models intended to help immigrants and others navigate fast-changing rules and regulations, complete complex forms, and prepare materials for human legal review, including appeals of detention or removal orders. These systems are not "virtual lawyers" in the full sense, but they are a part of a growing ecosystem of tools aimed at widening access to legal aid. Similar examples abound.

What will Kallibou and millions like him be able to accomplish when AI doesn't just translate languages or optimize routes, but fundamentally democratizes access to knowledge, capital, and opportunity? As we reach the near threshold where AI truly begins to reason with full and reflective coherence, what will be the first question we ask it?

I realize, of course, that this view of knowledge-driven AI is not universal. Rich Sutton, a computer scientist revered as the "father" of recursive learning—the means by which algorithms begin to teach themselves—is a much-quoted figure by those making the case that the value and contribution of human knowledge rapidly collapses as machines learn. His basic argument, made in a 2019 blog post ✷, that computer power beats human knowledge and learning, is now well-worn among evangelists for a material AI worldview. In his 2019 essay, "The Bitter Lesson," he wrote: "The biggest lesson that can be read from 70 years of AI research is that general methods that leverage computation are ultimately more effective—and more scalable—than human-designed, domain-specific solutions."

But his later and deeper reflection, in the same essay, usually escapes notice. It's worth a brief focus: "The actual contents of minds are tremendously, irredeemably complex; we should stop trying to find simple ways to think about the contents of minds, such as simple ways to think about space, objects, multiple agents, or symmetries. All these are part of the arbitrary, intrinsically-complex, outside world. They are not what should be built in, as their complexity is endless."

So, the question becomes: Just how will we embrace a sudden attachment to the forces of this endless complexity? What will our choice be? The answer may well redefine what abundance means for humanity itself.

Visions for the Future, and the Road to Get There

As AI pioneer Dario Amodei, founder and CEO of Anthropic, wrote in his seminal 2024 essay, "Machines of Loving Grace," which is essential reading: "It is critical to have a genuinely inspiring vision of the future, and not just a plan to fight fires." For me, that vision must include the Age of Abundance for All.

I would suggest that vision we can choose is one where AI and robots become the workforce, toiling in the fields, arc-welding in the factories, building affordable housing, and augmenting the power of health care workers—from home health aides, to synthetic biology researchers, to neurosurgeons. In support of that goal, some fifteen months after that original essay Amodei graced us with a second deeply insightful meditation on our AI moment in his essay, "The Adolescence of Technology," published in January 2026. An earnest and vulnerable fireside chat on the awe-inspiring hope and the terrifying risks of AI, it reads as the work of a deep thinker who has not retreated from optimism but has earned it by framing our technological ascent as a moral rite of passage that will test our institutions. "The years in front of us will be impossibly hard, asking more of us than we think we can give," Amodei wrote. "But in my time as a researcher, leader, and citizen, I have seen enough courage and nobility to believe that we can win—that when put in the darkest circumstances, humanity has a way of gathering, seemingly at the last minute, the strength and wisdom needed to prevail. We have no time to lose." I very much agree.

Or consider the vision and analogy of AI scholar Ksenia Se, founder and publisher of the AI newsletter, the *Turing Post*: "If we are bold enough to imagine it, it might look less like our society today, and more like something from the past: an aristocracy. Not an aristocracy of bloodlines, land, or oppression, but an aristocracy of the human spirit." Stripped of the context of feudalism, Se's framing is instructive:

"An abundance of resources, provided by the work of others, created the time and space for education, art, philosophy, political discourse, and self-cultivation. The aristocrat's purpose was not to earn a living, but to live a life of meaning, contribution, and intellectual expansion."

I'd add to that a commentary from AI scientist Roman Yampolskiy at the University of Buffalo. He suggests that a more urgent task than the much-debated case for a universal basic income (UBI) might well be the need to provide what he calls "UBM," for universal basic meaning. Yampolskiy effectively restates the thesis of psychiatrist and Holocaust survivor Viktor Frankl, author of *Man's Search for Meaning*, one of my top three books and one that all of humanity should read.

Other thoughtful visionaries leading the search for meaning in the coming metamorphosis, thinkers whom I admire and about whom I've written or spoken about extensively, include Google DeepMind co-founder, and now CEO of all of Microsoft's AI initiatives, Mustafa Suleyman; and Google DeepMind's other co-founder, and now CEO of all of Google's AI initiatives, Demis Hassabis, may become the first person to win two Nobel prizes since Linus Pauling. Former CEO of Google Eric Schmidt is a towering figure, and his work includes his book on AI, *Genesis*, co-authored with the late Henry Kissinger (it was Kissinger's final reflection on the future) is a gift to humanity.

Ray Kurzweil, the always insightful thinker, is behind the idea of a "Singularity" and is, along with Diamandis, the founder of Singularity University. In his latest work, *The Singularity is Nearer*, Kurzweil contends that we're sprinting toward a point when AI and human intelligence merge, ushering in what he calls "nonbiological immortality." In that book, which I reviewed as *The AI-Driven Universe a Blink of the Eye Away*, Kurzweil envisions a future where AI not only accelerates breakthroughs in healthcare, cognition, and longevity, but redefines what it means to be human.

Former OpenAI Chief Technology Officer Mira Murati, now founder and CEO of Thinking Machines Lab, ended a long stint in stealth in late 2025 when her company launched its first product, focused on refining and fine-tuning frontier models so their most powerful capabilities become more accessible, interpretable, and ultimately more beneficial to humanity.

Dr. Fei-Fei Li, the founder of Stanford's Center for Human-Centered AI, whose work helped define modern computer vision and whose moral clarity has consistently anchored AI to human values, is a national treasure. Her book, *The Worlds I See*, is—in a very different way—as powerful an immigrant's story of resilience and courage as is Kallibou's. I wrote of her amazing journey from China in my 2024 review, "Attention AI technologists, it's time to merge intentionality with philosophy, ethics, and law." Li, in *The Worlds I See: Curiosity, Exploration, and Discovery at the Dawn of AI*, interweaves her personal journey as an immigrant with the evolution of artificial intelligence. She argues that AI must remain human-centered. In that marvelous book, Dr. Li emphasizes the role of curiosity, inclusivity, and agency in ensuring that AI serves all of humanity, not just a privileged few.

We'll also discuss in detail the work of Reid Hoffman, both in coming chapters and in the "New Models for Humanity" section as Model 6, "The Philosopher-Technologist of the Emerging Age." He is the co-founder and former CEO of LinkedIn, whose most recent book, *Superagency—What Could Possibly Go Right with Our AI Future*, with Greg Beato, is among the most thoughtful treatises on AI ever written.

In his book, *The Coming Wave,* Suleyman argues that we stand at the edge of a new technological tsunami. I wrote about this in "A starting toolkit for humanity: Navigate the road to a future of AI-empowered Conscious Capitalism." Driven by AI and synthetic biology, the tsunami Suleyman describes promises "radical abundance" alongside risks that we must acknowledge and build strategies to avoid.

Another standout among researchers is my dear friend Byron Reese, an AI entrepreneur and author of four books including *We Are Agora—How Humanity Functions as a Single Superorganism That Shapes Our World and Our Future*. Agora, of course, was the ancient Greek marketplace where goods, ideas, and innovations were exchanged. Byron and I discussed his book in early 2025 on the *Austin Next* podcast, and later in the year as episode 12 on the *Love Conquers Fear* podcast. A planetary agora is humanity's destiny, Byron posits.

Byron is an outlier among AI researchers in the sense that he challenges the idea of AI ever reaching a consciousness comparable to that of humans, specifically artificial general intelligence (AGI). But his very human-faced vision of AI is that it can take us to an emergent era of "distributed superintelligence"—a superorganism—that allows humanity to function as a collective, not unlike the way bee hives or ant hills function as individuals acting in perfect, choreographed harmony. Unlike ants or bees, the goal is not a bigger hill or more honey, but the tools for all of us to collectively cure all disease, eliminate poverty, and defeat such existential threats as climate change.

"We can now face anything that fate can throw at us, and we can overcome it," Byron writes in *We Are Agora*. "But only if we can work together to do it. That's the trick." Byron's "trick," in fact, is the broad subject of this book, and my own life's work. This is the *intentionality* that we must bring to our development and use of AI, a subject central to the work and writing of Suleyman and Li.

Eric Schmidt is hardly sanguine about the challenges and risks of our AI future. In his book, *Genesis*, with Kissinger and also Craig Mundie, he explores manifold perils: machines misaligned with human values, authoritarian abuse of power, autonomous weapons without human decision-making, and critically, the danger in our current AI arms race with China. Importantly, Schmidt and his co-authors highlight what he calls "epistemic risk" in an overloaded information ecosystem, undermining public trust and shared reality—a focus of the work of Tristan Harris, whom you met in the last chapter.

But, in addition to his optimism that we can meet these challenges, if we are intentional, in a conversation with Diamandis on his *Moonshots* podcast in the summer of 2025, Schmidt shared his view that the imperative to collaborate may only come from some kind of attention-riveting "national emergency." But the payoff for thoughtful collaboration will be a "polymath in your pocket," a genius smarter than any who ever lived, who will be with you always. As a step toward enhancing democracy worldwide, he envisions a product that "teaches every human who wants to be taught, in their own language, in a gamified way, the stuff they need to know to be a great citizen in their country." In the future envisioned by Schmidt: "You will apply AI to what fascinates you. Education then becomes a response to the call to purpose—whether that be climate, or materials science, or energy systems." Imagine what that could mean for Kallibou and Chad.

A further example of what I mean by intentionality is a radical proposal from another good friend, Carla Piñeyro Sublett, the former chief marketing officer for IBM, who now consults with CEOs and executives around the world on collaborative leadership. We spoke on her *Let's Go There* podcast, which re-aired as episode 7 on the *Love Conquers Fear* podcast.

I'm sure readers here are familiar with the debates around data privacy and data portability, which has led to a still nascent movement in a number of countries and U.S. states aiming to allow each of us to "own" the vast amounts of data harvested from our interactions with websites, social media, streaming services, or chatbots. Carla is a leader in an emergent movement supported by the likes of World Wide Web inventor Tim Berners-Lee that would take this one step further. Carla foresees a world where we own our own algorithms, the mathematical formulations that allow Netflix to suggest your next movie or your browser to serve up ads based on the search you just made for a replacement lawnmower.

Imagine the world that Carla envisions in which you control the personalized logic and decision-making rules that process that data—the filters, preferences, predictive models, and AI "settings" that shape how systems interact with you. Later, we'll discuss government, politics, regulation, and policy. But imagine if behind-the-curve lawmakers could move beyond imposing well-intentioned but fear-based constraints on technology to supporting and hammering out ideas such as Carla's. This would be a fundamental power shift toward the Age of Abundance for All in the digital economy, and we could do it with AI.

I've written a great deal about education and AI, a subject we'll explore much further in Chapter 6, "Education." My essential argument is that our education model is linear, in an increasingly non-linear world, a situation that prompted me to invent a new word for the dilemma: *linearalism*, which I coined for my essay, "Youth Will Lead Us to a Much Needed, New 'Learning Ecology." So I'm heartened to see in the AI discourse the rediscovery of a book first published in 1970, *Deschooling*

Society , by the late Austrian priest and polymath Ivan Illich.

Illich's idea was to get rid of institutionalized education—the K–12 and bachelor's, master's, and doctorates—with which we're familiar. He proposed replacing them with a concept of "learning webs," where networks of people, skills, and resources would enable the exchange of skills, knowledge, and mentoring. We are already seeing the early shoots of this blossoming metamorphosis of learning in such examples as Sal Khan's Khan Academy, offering countless how-to videos on YouTube or the infinite number of online classes such as those offered by Coursera or MasterClass, though these are still based on fixed curricula—again, linear. AI will usher in the post-linearalist era of learning.

Were Illich, who passed in 2002, to return today, I'm quite sure he'd see AI as the missing infrastructure of his idea. Imagine if online learning were entirely customized, even powered by Carla's personalized algorithms. With an AI tutor trained on your personal skill level and learning aspirations, the acquisition of knowledge will increase at warp speed. In Chad, students with an entry-level smartphone would have all the knowledge resources made accessible through a tutor fully versed in their skill level and communicating in their native Ngambay. Trading insights and ideas through instant translation, on an equal footing with peers in Cambridge, Massachusetts, or Oxford, England, the Chadian student would be an early pioneer in Byron's superorganism, the Agora. And perhaps these distributed peers would collaborate to confront challenges, from climate change to the preservation of the Lake Chad fisheries, with solutions that have so far eluded them.

As I suggested above, Hassabis, CEO of Google DeepMind, which combines Google's core AI initiatives, is in a category all of his own, and I wasn't kidding when I speculated he'll be the sixth person in history to win two Nobel prizes (the first was Marie Curie, who won in both chemistry and physics, which Hassabis could well repeat). He won his first Nobel in chemistry last year, with John M. Jumper and David Baker, for his work on AlphaFold, which cracked the code of protein structures. I also see him as a candidate for the Nobel Peace Prize, for leading awareness that abundance must be for all.

In 2025, Hassabis issued an intriguing call for like-minded nations to form a powerful coalition in AI. As a counter to the expanding U.S.–China AI arms race, he articulated a vision for a third bloc, capable of shaping the future of human history's most transformative technology with a voice rooted in shared democratic values: "I think the UK and France and a few other like-minded countries, perhaps like Canada and Switzerland, if we band together with the same values, can form a counterweight to the two global superpowers and influence the debate at the global stage. I think that's very important to have that voice, given that this technology is going to affect everyone in the world."

That's a quick overview of the stakes. And they are before us as the pace of this metamorphosis is hard to imagine or describe. AI's ubiquitous use and influence is like a river bursting its banks. It's not something we can control or easily channel, but it is a current whose power we can harness and yoke to the reach of the Age of Abundance for All.

We're on the Clock to Make Our Choice

The best description of this velocity is to be found in an interesting, much-debated, and widely discussed report published in the spring of 2025, *AI 2027*, which forecasts the power of AGI—the moment when AI systems are capable of reasoning, planning, and learning as well or better than humans, will be here by 2027. Researched and produced by the AI Futures Project, a nonprofit think tank led by former OpenAI research scientist Daniel Kokotajlo, it presents a rigorous, month-by-month forecast of rapid AI evolution driven by self-improving AI agents. This requires vast increases in computational power enabled by massive data centers that are already using 1.5 percent of the world's energy supply, with consumption on track to double by 2030. That's about 470 terra-watts of energy, enough to power every household in New York, Texas, Florida, and California combined. AI workloads take about 20 percent of that data center capacity, a share set to double within the year as it competes for electricity with cloud services and crypto currency mining.

Under one scenario in *AI 2027*, the world navigates the emergence of powerful AI agents with foresight, cooperation, and successful alignment. AGI does arrive, but it is controlled, understandable, and beneficial. Humanity will quickly evolve in that scenario from AGI to ASI (Artificial Super Intelligence) and then GLI (God-Like Intelligence), only limited by the laws of physics. But, in the dystopian scenario, *AI 2027* lays out chilling potential trajectories, from disruptive breakthroughs to misaligned superintelligence collapsing the U.S. government and creating chaos throughout the world.

The upshot of the 2027 simulation, however, is the argument I am waging here. The future of AI is not predetermined by technology alone. It depends on the decisions we make, about openness, governance, incentives, and responsibility. Whether we enter an unprecedented age of abundance or a preventable tragedy, the choice is ours. The difference lies in how wisely we collaborate. God gave us free will, and it is time for us to use it.

I'm 95 percent certain we will make it to the Age of Abundance for All, and of course it bothers me that there is a 5 percent chance that we'll extinct ourselves. That's why I launched the *Love Conquers Fear* podcast in September of 2025, to explore how we'll make it with some of the leading technologists, entrepreneurs, scientists, and spiritual teachers.

Capacity versus Agency: The Path to Superagency

As we stand on the threshold of this metamorphosis, we must distinguish between two critical forces: *capacity* and *agency*. The family of exponential technologies we have discussed—the Superfecta—provides us with an unprecedented capacity for abundance. We now have the "polymath in our pocket" and the computational power to solve once-intractable problems like the drying of Lake Chad or the collapse of global wetlands.

However, capacity alone is a dormant engine. As Hoffman argues in his concept of "Superagency," technology provides the tools, but humans must provide the agency to direct those tools toward flourishing.

- **Capacity** is the existence of an LLM that can draft a legal appeal for an immigrant.

- **Agency** is the intentionality of the Stanford engineers who chose to build it.
- **Superagency** is what happens when we use these tools to function as a "collective superorganism," moving beyond the fear of the tech titans to solve existential threats like climate change together.

An Oasis of the Spirit

As my ride with Kallibou came to an end at the airport, the sunrise was just beginning to hit the glass of the terminal. I realized that my friend from Chad is not just a passenger in this technological revolution; he is its most important pilot.

His journey—six years of separation, supporting a family across an ocean, and driving through the Austin night—is the ultimate expression of human agency. If we provide him with the capacity of the Age of Abundance for All, his agency will do the rest.

I imagine Kallibou not just saving for a plane ticket, but using a "polymath in his pocket" to design the very water pipelines and solar arrays that will turn Chad into an oasis. In this vision, he doesn't need to bring his family to the lamp of the Statue of Liberty; he brings that light home.

Let's now shift to Part Two of this book on "Application" to explore and project solutions to the biggest problems that face humanity today.

However bright this vision of abundance, technology alone cannot get this job done. Instead, the vision will require all of us. So let's get busy.
—Peter Diamandis

APPLICATION

CHAPTER 3
GEOPOLITICS

*The salvation of this human world lies
nowhere else than in the human heart, in the
human power to reflect, in human meekness
and human responsibility.*

**—Vaclav Havel, author, poet,
winner of Gandhi Peace Prize, and
former President of the Czech Republic**

My friend and co-entrepreneur Pablos Holman makes an insightful case in his new book *Deep Future* , which I reviewed as "From Mosquito Lasers to Self-Healing Concrete: An AI Genius' Blueprint for Abundance." The argument is that we construct our reality through modeling: A child organizes her dolls in a make-believe school as prep for her first day in class—a model. An accountant models a business with a spreadsheet. Starbucks models imaginary lattes at various price points before a product launch. DeepMind's AlphaGo modeled playing the Chinese board game Go billions of times to the point where it beat the grand human master of the ancient game.

Think through this idea of modelling in terms of your own life and work. We understand the world through the models we construct and hold in our minds: "TV shows, maps, novels, board games, video games—all [metaphorical] models," Pablos writes. They are critical. Pablos' innovations based on this precept include one that helped successfully eradicate polio in Nigeria.

As we move from a society organized around scarcity toward one built upon abundance, we are struggling to envision new models of sovereignty, governance, and democratic institutions. No one described this dilemma better than the American Founding Father Thomas Jefferson, who wrote in an 1816 letter:

> I am not an advocate for frequent changes in laws and constitutions. I think moderate imperfections had better be borne with; because, when once known, we accommodate ourselves to them, and find practical means of correcting their ill effects. But I know also, that laws and institutions must go hand in hand with the progress of the human mind. As that becomes more developed, more enlightened, as new discoveries are made, new truths disclosed, and manners and opinions change with the change of circumstances, institutions must advance also, and keep pace with the times. We might as well require a man to wear still the coat which fitted him when a boy, as civilized society to remain ever under the regimen of their barbarous ancestors.

I love this quote and the intent of it. But note that in Part Three, "Transformation," we'll be talking more about how the stereotype of "barbarous ancestors" was used to suppress some of the most ancient technology that humanity had for transformation. Jefferson was, of course, a person of his time and had a few other sins we know about. But let's keep in mind that our own primitive views of Indigenous people were part of what allowed capitalism to trounce over native Americans in the rush to "progress" and "enlightenment." When Thomas Jefferson penned these words, he was offering a warning that we have largely ignored: the structures we use to govern ourselves are not eternal truths; they are "models" designed for a specific time and a set of problems.

Please understand, I'm a huge fan of American Revolutionary history; the genius of the American Founding Fathers and the wisdom of Jefferson's letter resonate with today's reality in many ways. But an all-but-lost chapter of our own geopolitical history is the contribution of Indigenous political thought to our own. I recommend a riveting book by David Graeber and David Wengrow titled, *The Dawn of Everything*. It highlights how the Iroquois, or Haudenosaunee, Confederacy provided a living model of political organization that directly influenced Enlightenment thinkers and, eventually, the American Founding Fathers. Their model was built on:

- **Decentralized Power:** Six distinct nations were acting as one while maintaining local autonomy.
- **Long-term Thinking:** Decisions were made with the "seventh generation" in mind.
- **Consensus-Based Stability:** The system was designed to absorb shock and conflict through dialogue rather than force.

The Haudenosaunee Confederacy presented a real-world example of a democratic confederation that inspired strands of thought that fed into the framing of the U.S. Constitution—an important lesson to remember as we draw on other ancient beliefs and wisdom as we approach humanity's spiritual "bend in the river." The full history of the Haudenosaunee's remarkable contributions to our nation's founding ethos, including their influence on Benjamin Franklin, is explored among the New Models for Humanity as, "How America's Founding Was Shaped by Indigenous Wisdom."

The Obsolete Chessboard

Back to our day, fast-forward through two centuries of Jefferson's "progress of the human mind" and many models remained unchanged. None more so than in our approach to geopolitics, where the most ubiquitous mental model we all use for purposes of conceptualization is chess—a game invented for the realities and imperatives of the seventh century. "For centuries, chess has been a metaphor for war in myth and in literature," wrote Harvard's Kenneth Rogoff. How many times have you encountered language like this in a recent piece in the *New York Times* on a proposed Middle East deal: "It rearranges the entire chess board?" When that master of geopolitical strategy, the late Zbigniew Brzezinski, published his 1997 book on the future of diplomacy, he titled it *The Grand Chessboard: American Primacy and Its Geostrategic Imperatives.*

The irony of this inversion of reality as game is palpable. It's akin to "how to think outside the box" became one of the most inside-the-box things one can say. Only in this case, the scale of irony is planetary. Since metaphors aren't just linguistic flourishes but the basic architecture of human cognition—we literally cannot think without them—this matters profoundly. The geopolitics-as-chess paradigm has become as obsolete as the feudal world that created it.

Reframing our worldview around this insight illustrates how many of our problems today—from wars in so many places, to incoherent politics, to "crises" in education, security, and climate—are really symptoms of a meta-problem: Our fast-changing world is becoming one for which we lack effective models. What we need, as we approach humanity's "bend in the river" that we explored in Chapter 1, "Conquering Fear", is what I'll call a "geopolitics of unity." Here I mean unity in the sense of Martin Luther King Jr.'s beloved community, Gandhi's satyagraha (soul force), or Frederick Douglass' moral suasion. It's not going to be an easy journey, and before the bend, we must cross through the whitewater of fear that is being churned by this outdated worldview—the model of chess that insidiously reflects the mindset of scarcity.

Toward this new reality, I suggest that we do actually have an updated model. It's still imperfect but better, and it's one you've probably heard of: *Raumschach,* or "space chess," invented in 1907 by Ferdinand Maack. It sought to add to chess the changed reality of politics, diplomacy, and war with the innovations of airplanes and submarines. Until Maack's age, the spatial organization of geopolitics was horizontal. With these new weapons of war, it became vertical as well. In his version of the board game, with multiple boards stacked three or more high, players must learn to checkmate vertically as well as horizontally.

Image 3.1: 3D Chess, reimagined with AI.

Beyond a few *Star Trek* episodes and nerdy students like myself, 3D chess never caught on as a pastime, let alone as a metaphor. And I don't suggest it is complete—all models are imprecise heuristics that move us closer to understanding—but it represents a start. Our geopolitical chessboard remains two-dimensional in our perceptions: good guys, bad guys; allies, enemies; spheres of influence; balance of power.

We still think in terms of discrete nation-states maneuvering for advantage on a flat board, as if the world were still organized around the territorial ambitions of medieval kingdoms—a paradigm of hierarchical power reflective of seventh century India's Pushyabhuti Dynasty where chess was invented, with rival kingdoms pitting their four divisions in contest on a single plane or dimension: infantry, cavalry, elephantry, and chariotry. The metaphor assumes clear boundaries, identifiable opponents, and conflicts that end with someone's surrender.

Today we've added not just submarines and airplanes to the landscape, but nuclear weapons, supersonic missiles, cyberwar, drones, computational propaganda, all now with AI woven through the architecture of communications, command, and control. We haven't really understood how geopolitics became three-dimensional, and now we're grappling with its fourth dimension—the Superfecta, whose utility and scope are accelerating at exponential speed.

When Seventeenth-Century Models Meet Twenty-First-Century Realities

Our fundamental model remains the nation-state—a seventeenth-century concept born from the Treaty of Westphalia in 1648. You could say we installed a new geopolitical operating system—Westphalia 1.0. The seventeenth-century nation-state was a brilliant "OS" for its time, designed to manage national systems of education, economic management, and social welfare. We then locked it into place with the post-World War II "rules-based order," modeled on new institutions like the United Nations, the World Bank, and the International Monetary Fund.

If this is our planetary operating system, the need for an upgrade is inescapable. The looming challenge of climate change represents the most obvious example—a challenge toward which nations accustomed to acting in their own self-interest are proving profoundly ineffectual. National borders have proven meaningless against viruses, as we learned during the Covid-19 pandemic, and as we are now in danger of forgetting. Above us, space grows more dangerous to travel and explore as more than 6,600 tons of debris race around Earth at 18,000 miles per hour—while the deployment of private mega-constellations accelerates. Regulators have cleared SpaceX's Starlink to expand its Gen2 constellation to 15,000 satellites, and Musk's company has even filed a proposal for up to one million "orbital data center" satellites, in a crowded commons still lacking enforceable global governance.

Meanwhile, here on Earth, the Great Pacific Garbage Patch—1.8 trillion pieces of plastic covering an area twice the size of Texas—floats with no state or authority claiming responsibility for the problem. The need for the United States and other developed nations to control borders remains urgent, but how effective can any effort be when we lack models to cope with a highly mobile worldwide refugee population greater than Japan's entire citizenry of 120 million?

Or consider this: As of early 2026, the combined market capitalization of the "Magnificent Seven"—Apple, Microsoft, Alphabet, Amazon, Nvidia, Meta, and Tesla—surpassed roughly $20.9 trillion. That figure stands above the annual GDP of nearly every nation, except the United States and China, and rivals or exceeds the GDP of the entire twenty-seven nations of the European Union.

So, just who runs the world? Scott Galloway, mentioned earlier, author of *The Four—The Hidden DNA of Amazon, Apple, Facebook, and Google*, was more right than we fully understood when he published that stirring book in 2017. He notes that his criticism of Big Tech's outsized clout and bullying behavior is implicitly acknowledged even by the companies' leaders themselves: "The CEOs sit and listen to my talks and smile," Galloway wrote back then. "It's the smile of poker players holding aces. And every one of those aces is data. In the last decade, the world's most important companies have become experts in data—its capture, its analytics, and its use." Can those nominally in charge, national governments, close this gap in the mastery of data, which is clearly the feedstock of AI and root of copyright battles today?

As we stand at the choice point of 2026, it is clear that we can no longer run twenty-first-century complexity on seventeenth-century code. We are currently experiencing what happens when a "model" reaches its breaking point. The "glitches" we see in our daily headlines—the inability to coordinate on global health, the friction of digital borders, and the failure of traditional institutions to regulate borderless technology—are not just policy errors. They are system crashes.

The inability of our institutions to cope with this inordinate power was captured insightfully in the 2024 book *Tech Coup* by Marietje Schaake, a former Dutch member of the European Parliament now at Stanford University's Institute for Human-Centered AI, an institution about which I've written. Her summation of the governance challenge is sobering.

On one side of the Atlantic, American lawmakers embarrassed themselves in an intended grilling of Meta CEO Mark Zuckerberg, displaying to the world their ignorance of the very company and industry they're meant to regulate. "Few in the Senate seemed capable of logging into Facebook," Schaake observes. On the other side of the Atlantic, starstruck European MPs also failed to meaningfully question Zuckerberg, and then rushed from the dais to snap selfies with the tech titan as he cut short his testimony. Elsewhere, Schaake documents India's nightmarish national biometric data collection, China's perfection of the Orwellian superstate, France's culpability in a Kenyan election cancelled amid technological fiasco and incompetence, and the stealth acquisition of land and resources for electricity-gorging data centers through shell companies around the world.

Thomas Jefferson understood this. When he spoke of the "regimen of our barbarous ancestors," he wasn't just being provocative; he was acknowledging that the models of the past have expiration dates. The pattern emerges clearly: seventeenth-century institutions confronting twenty-first-century realities, with predictably inadequate results.

The task ahead is not to "improve" the old OS or patch the existing code. Our "barbarous ancestors" gave us a coat that no longer fits the stature of a global, interconnected species. We must transition from trying to fix a failing seventeenth-century model to recognizing that it has become an obsolete operating system, unsuited for the high-velocity world we now inhabit. As we look at the recent system crashes, the question is not how to return to the seventeenth century, but how to reclaim the wisdom of those "ancestors" who understood that true governance isn't about control—it's about the flow of peace and the sustainability of the collective.

The Challenge of Complexity

One who has wrestled with these questions in profound if still-evolving ways is Cameroonian postcolonial theorist and philosopher Achille Mbembe. He advances a planetary model of equitable governance that decenters the nation-state in favor of shared stewardship of the Earth as a common habitat, anchored in new universal rights that extend to humans and nonhumans alike. He asks: "Can we rely on infrastructures that have, to some extent, contributed to turning the world into a burning house? Can we rely on them to learn how to inhabit the planet anew, how to share it as equitably as possible? To foster a new consciousness that gives ample space to notions of bio-symbiosis—life in symbiosis with humans and nonhumans?"

These are the essential questions. Yet they are ones that, in my view, we can't answer—not without new models of thought and action and a new, post-scarcity consciousness for this moment. Developing these new models is complex, to say the least, and complexity itself is a concept we fail to wholly grasp. To explain, imagine I ask you to describe the opposite of simplicity. Your reasonable sounding answer might well be complexity. That answer itself is a flawed model. The opposite of simplicity is better described as complicatedness—problems that can be intricate yet remain largely predictable and decomposable through linear approaches. Complexity, however, operates on an entirely different spectrum. Complex problems are adaptive and non-linear, producing emergent outcomes that can't be solved through traditional planning or analysis. They demand fundamentally different approaches than our toolkit of complicated solutions.

As I wrote some years ago in an essay, "Data is from Mars. Data Science is from Venus," we attempt to convert complex challenges into their merely complicated components. This is the history of human endeavor. On our journey to modernity we have converted forests to orchards, wild jungles to monoculture farms, fisheries to aquaculture, and rivers conveying commerce into dams transforming kinetic energy into electricity. Complex problems—of unknowns, of infinitely interrelated factors—don't yield to our toolkit of complicated solutions, no matter how intricate. Yet, even as we recognize the inadequacy of our current models, we find ourselves doubling down on the very frameworks we must escape.

The U.S. administration's $500 billion Stargate AI Initiative, announced on January 21, 2025, exemplifies this perfectly. It's a massive national effort to beat China in developing artificial general intelligence (AGI), the ambiguous goal of creating AIs that are as smart as any living human today in any field. The timing proved striking: just one day earlier, Chinese startup DeepSeek had unveiled its open-source R1 model with reasoning performance matching OpenAI's models, developed at a fraction of the cost. Whether coincidence or not, the juxtaposition reinforced the narrative we've organized around this civilizational challenge—yet another zero-sum contest between nation-states. It's a subject I discussed with entrepreneur and geopolitical analyst Pippa Malmgren on episode 35 of the *Love Conquers Fear* podcast, as our response continues to be dominated by export controls, sanctions, and the familiar rhetoric of technological supremacy, which she calls "our path to extinction." Most concerning is that we have locked ourselves into a manic race to create AGI and prospectively consciousness (with recursive self-improvement) when we've yet to fully understand what consciousness is.

This is not an argument against progress. To be clear: I advocate for developing AI as rapidly as possible, but the challenge lies in maintaining intentionality. We must remain focused and clear about our destination. While I opposed the March 2023 calls for an AI development moratorium as both misguided and unrealistic, I understand the underlying concerns of its advocates. As we advance beyond the capabilities of GPT-5, we face a collective humanitarian challenge of ensuring these increasingly powerful models serve humanity through robust AI safety efforts—the very concerns that helped drive AI pioneers like Mira Murati, Ilya Sutskever, and Dario Amodei to leave OpenAI at different moments and establish their own companies, and that also served as a major motivation for writing this book. Others of like mind in Silicon Valley include LinkedIn co-founder Reid Hoffman and Stanford professor Fei-Fei Li, whose work we'll explore in further detail. As discussed in the last chapter, Amodei has written eloquently on the reality of AI risks and our responsibility to mitigate them as we move toward the promise of AI technology, most recently in his essay "The Adolescence of Technology," published in early 2026.

This AI arms race represents precisely the kind of binary, chessboard thinking that our complex moment demands as we transcend two dimensional geopolitics. The development of AGI—technologies that will redefine labor, identity, and reshape our moral frameworks in ways we can barely imagine—requires deliberate, ethical navigation that serves humanity's collective benefit, not just national advantage. Adapting our institutions to harness AI's transformative potential in healthcare, education, energy, and beyond will be the greatest challenge of our time.

One much-quoted critic of our current approach to AI is Mo Gawdat, the former chief business officer for Google X, Google's advanced innovation arm. Gawdat has argued that unpreparedness for AGI could plunge the world into a fifteen-year period of geopolitical chaos and societal meltdown, what he calls a "hellish" transition of economic disruption and inequality if political pressures are allowed to spiral. I don't share Gawdat's grim outlook, although it is a *possibility* if we don't really guide ourselves with intentionality, but I do share his hope that this could be transformed into a what he calls a "heavenly" outcome through proactive governance, ethical foresight, and equitable deployment of AI. I'll be discussing that more in Part Three, "Transformation."

Gawdat is correct that in so many ways we're approaching this challenge with the same dichotomous Manichean mindset that emerged from the Cold War, the same competitive nationalism that is proving so inadequate for addressing climate change, space debris, or ocean plastic. The manic race mentality—us versus them, first-place or failure—reflects the very two-dimensional thinking that our four-dimensional world has rendered obsolete. We need new models of cooperation and governance, not faster versions of old ones.

The policy landscape for meeting AI's geopolitical challenges is moving rapidly—and nowhere more visibly than in the weeks just before this book went to press. In an announcement that drew immediate attention across the AI world, on April 6, 2026, OpenAI CEO Sam Altman released a white paper, *Industrial Policy for the Intelligence Age: Ideas to Keep People First*. In many ways, it echoes the concerns of Anthropic CEO Dario Amodei in his essay almost three months earlier, *The Adolescence of Technology*, that we explored in Chapter 2, "Abundance."

Like Amodei's essay, the OpenAI white paper acknowledges similar categories of risk—economic disruption, misaligned systems, and concentration of power. But the document goes further with concrete recommendations, calling for AI data centers to generate local jobs and tax revenue and pay their own way on energy. It suggests international auditing standards designed for global adoption, and a global network of AI safety institutes—essentially a kind of NATO or UN agency-equivalent for AI risk. Among ideas to help society navigate superintelligence and its societal impacts, the paper asks the federal government to tax AI-driven profits, create a wealth fund, implement a four-day workweek, and more.

It also very specifically touches upon a topic we'll explore deeply in Chapter 9, "Capitalism" with its recommendation that all frontier AI companies adopt the model of the Public Benefit Corporation, which both OpenAI and Anthropic have done.

In a candid interview with Axios's Mike Allen, Altman explained the urgency behind the white paper, pushing back directly against the view that governments should stay out of AI regulation. He called instead for deep public-private partnership—an approach he said is drawing provisional support from both senior Republicans and Democrats. And he issued a stark warning: rogue AI systems capable of designing devastating cyberattacks or engineering weapons from novel biological pathogens are not a distant threat. The arrival of AI superintelligence, he told Allen, "is so close, so mind-bending, so disruptive that America needs a new social contract—on the scale of the Progressive Era in the early 1900s, and the New Deal during the Great Depression."

Three Models for What's Next

Other new models of thought are also emerging on the horizon. They remain incomplete works-in-progress, but they hold promise to calm the whitewater rapids in humanity's current of consciousness as we ready to navigate that bend in humanity's river that extends beyond our planet to the stars and cosmos beyond. None represent our goal of a "geopolitics of unity," but all are incremental vehicles that can help us get there.

The Technopolar Reality:

Back in 2021, Ian Bremmer, founder of the geopolitical consultancy Eurasia Group, coined the phrase "technopolar moment." Recognizing that our chess-board metaphors had become not just obsolete but dangerously misleading, Bremmer observed what Jefferson recognized: that our institutions had failed to keep pace with reality. As mentioned above, a handful of large technology companies now operate as quasi-sovereign actors, exercising state-like authority over the digital realm and increasingly shaping geopolitics, economics, security, and social life through control of data, servers, platforms, and algorithms.

This represents a fundamental shift in how we must understand power itself. Sovereignty in this arena is no longer defined by territory or military force but by governance of digital infrastructures and standards that now mediate the daily lives of billions and underpin critical national capabilities. We've moved from the two-dimensional world of bordered nation-states to something far more complex—a reality where neither traditional states nor private firms alone possess the capability to govern effectively.

Bremmer's insight cuts to the heart of our institutional crisis: effective governance now requires formal roles for major tech firms in global rulemaking, given their practical sovereignty over digital systems and their ability to shape outcomes beyond any single jurisdiction. The defining political challenge becomes managing this fusion of tech and state power while preserving accountability, rights, and democratic control in a world that has outgrown our seventeenth-century frameworks.

Yes, this framework raises questions about democratic legitimacy.

If tech companies wield quasi-sovereign power, who elected them? How do we ensure accountability when private entities control infrastructure that governments themselves depend upon? Bremmer's technopolar model accurately describes our reality but offers few answers about how to democratically govern it.

Network States—A Radical Alternative:

Balaji Srinivasan, a prominent entrepreneur and technologist, proposes a more radical solution in his book *The Network State.* While I don't agree with all of his arguments, the former Coinbase CTO offers several intriguing ideas. Srinivasan argues that we don't need to reform our obsolete institutions, we can *transcend* them through what he calls "network states."

Srinivasan's vision flips the traditional sequence of state formation on its head. Instead of "land first," he proposes "community first," building highly aligned online communities around shared values, then using blockchain technology to create verifiable metrics of membership, economic output, and collective action. These digital communities would eventually crowdfund physical "archipelagos" of real-world properties, provide their own services, and seek diplomatic recognition from existing nations.

His core insight echoes Jefferson's but reaches a more radical conclusion: If institutions must evolve with human progress, why not build entirely new ones? The internet, he argues, has "unbundled identity and coordination from geography," making it possible to form political communities globally first, prove their viability with data, and only then negotiate for territory and recognition. In Srinivasan's vision, networks become the new basis of sovereignty—opt-in, measurable, transnational, and economically aligned rather than geographically constrained.

It's a compelling model that moves far beyond our two-dimensional chess thinking. It turns on imagined, digital communities that I've called "epistomes" and others have called "epistemic spaces." Rather than trying to coordinate among 190+ nation-states with conflicting interests, Srinivasan envisions governance by alignment and choice—people joining communities that share their values and prove their effectiveness through transparent metrics.

This model too faces enormous practical hurdles. Existing nations are unlikely to grant diplomatic recognition to competing sovereignties, especially those that might undermine tax bases or regulatory authority. The approach could exacerbate inequality by allowing the wealthy and technologically sophisticated to opt out of shared civic obligations—effectively creating premium governance for those who can afford it while leaving others stuck with deteriorating legacy institutions. And the emphasis on "highly aligned" communities raises questions about diversity, dissent, and minority rights within these new polities.

World Positive Capitalism—Reforming the System:

Even capitalism will require fundamental rethinking. I've long been drawn to the conscious capitalism and to the B Corporation and Public Benefit Corporation movements. My most recent company, data.world, was a pioneer in the B Corp movement for technology companies. Public Benefit Corporations legally mandate responsibility to community and society, not just to shareholders. I've also served on the board of the non-profit Conscious Capitalism with its co-founder and my good friend John Mackey, best known as the co-founder and former long-time CEO of Whole Foods. These models represent the kind of institutional evolution our moment demands.

While Srinivasan envisions escape from legacy institutions, another approach seeks to build on this movement to further transform capitalism itself from within. Vishal Vasishth, a good friend and co-founder of Obvious Ventures, offers what he calls "World Positive Capitalism" —a framework that acknowledges capitalism's contradictions while proposing systemic solutions rather than wholesale replacement. We'll discuss Vishal and his work in more detail in Chapter 9, "Capitalism."

Vishal's diagnosis aligns with our institutional crisis: Traditional capitalism has operated extractively, taking from natural resources and communities without considering long-term consequences. Even well-intentioned corporate efforts like Environmental, Social, and Governance (ESG) initiatives focus more on minimizing harm than maximizing regenerative potential. His proposed alternative rests on three interconnected pillars—planetary health, human health, and economic health—designed to create what he calls "compounding benefits" rather than zero-sum extraction.

The elegant model represents an important dimensional shift from our binary thinking. Instead of the traditional either/or choice between profits and purpose, Vishal argues for abundance-oriented systems that generate large profits *as a result of* positive impact. Companies like Tesla didn't just build electric cars—they redefined what clean energy transportation could be, forcing entire industries to follow. This approach moves beyond the chess-board mentality of winners and losers toward what he terms "exponential progress."

Vishal's vision particularly resonates with our complexity framework. Rather than optimizing for simple metrics like quarterly GDP growth, he envisions a scorecard to track multidimensional wellness—health, education, environmental sustainability, and social resilience. It's exactly the kind of non-linear measurement system our complex world demands.

Again, of course, questions remain about scale and implementation. Vishal's examples largely come from venture-backed startups serving affluent markets. Can World Positive Capitalism work for extractive industries, manufacturing, or services serving lower-income populations? How do you measure and incentivize "flourishing" across vastly different cultural contexts? And who decides what constitutes positive impact—investors, communities, or some combination?

Planetary Governance—The Systems View:

The most comprehensive attempt to address these dimensional challenges comes from Jonathan S. Blake and Nils Gilman in their 2024 book *Children of a Modest Star.* Building on the insight of philosopher Mbembe, they frame our moment as the "third de-centering" of human consciousness. This follows the first de-centering when Copernicus moved Earth from the center of the universe, followed by Charles Darwin, who displaced humans from the pinnacle of creation. This third shift forces us to recognize that we're not separate from planetary systems but embedded within them, a core belief of mine.

This transformation also reflects the thinking of another fine mind, that of LinkedIn co-founder and investor Reid Hoffman who emphasizes humanity's evolution from *homo sapiens* to *homo techne* throughout his work. This is the recognition that we are fundamentally shaped by the tools we create, from the first stone axes 300,000 years ago to today's AI systems. Blake and Gilman add to this insight, identifying what they call the emerging "Sensorium," a planetary array of instruments, algorithms, and integrated computer systems that create a fusion of machine and human intelligence. This technological convergence makes visible our deep entanglement with Earth's life-support systems, from ocean currents to atmospheric chemistry. We can now see, in real-time, how our actions ripple through interconnected planetary networks.

Blake and Gilman's diagnosis aligns with our institutional obsolescence: Governance models designed for separate, sovereign territories cannot address challenges that ignore borders entirely. Climate change, pandemics, space debris, ocean plastic—these operate according to planetary physics, not political geography. Even well-intentioned efforts like the Amazon Cooperation Treaty Organization to save the world's largest rainforest fail because "national resistance to delegating authority and resources, and claims to absolute sovereignty prevent integrated governance."

A better response? Blake and Gilman argue we need a shift from "global" thinking, still filtered through national interests, to genuinely "planetary" governance—what I call the geopolitics of unity. This includes non-human entities and natural systems in political representation. It's a radical reimagining that moves beyond both traditional nation-state frameworks and Silicon Valley's techno-libertarian alternatives toward something entirely new—governance by the logic of planetary systems themselves.

All these visions are compelling, and deserve our attention. But they of course raise fundamental challenges about human agency and democratic participation. How do you represent the interests of forests, oceans, or the atmosphere in political processes? Who speaks for these systems, and how do we ensure such representation doesn't become another form of technocratic rule by experts? And most importantly, how would these approaches conquer our fundamental challenge—fear.

A Model for the Geopolitics of Unity

Pablos Holman reminds us that models aren't just abstract concepts—they determine what becomes possible. And this is why our ultimate model must go beyond, to a geopolitics of unity and, ultimately, love.

Do I have a grand design for a "geopolitics of unity"? I'm trying, including in many discussions on my podcast *Love Conquers Fear*. This is something we must create together. One model that recently made my soul sing is a microcosm of inclusiveness, a project that conquered fear and one that must be part of our inspiration.

That model is a newly concluded project in Los Angeles' troubled neighborhood of Watts, long one of the most violent communities in America. Shortly after the pandemic, entrepreneur and Henry Crown Fellow, Michael Soenen , who lives in L.A., had an idea. He distributed dozens of iPhones to rival Crips and Bloods gang members, to victims of crime, to family members of victims, and even to interested—if skeptical—police. The request was simple, just film their difficult lives. Some two years later, with hundreds of hours of video in hand that were entirely filmed by the citizens of Watts, Soenen's team edited the archive into *Nothing to See Here: Watts* , a compelling documentary created by the community. It led to an historic meeting of former gang rivals, police, and the victims of crime. The mutual respect and empathy the initiative nurtured led to a 90 percent reduction in homicides and other violence. I had the honor of witnessing the film's premier at the 2025 Resnick Aspen Action Forum. Attended by gang rivals, victims, students, activists, and police, the film was one of the most moving I've ever seen. Afterward, all of them went on stage to the tune of U2's song "Where the Streets Have No Name," sung by the Watts' community choir, and hugged one another in a beautiful display of unity. There was not a dry eye in the house, including mine and our youngest child Yuzu's.

Yes, the coming transformation of our consciousness and our society will require leaders of all types. And those leaders will emerge from surprising places, as has always been the case—from America's Founding Fathers, to Mahatma Gandhi, to Nelson Mandela, or Mother Teresa. But the real leaders in this transformation will be us, Team Humanity. We must be the narrators of this coming story, and I dive into that in Part Three, "Transformation."

Image 3.2: Onetime gang rivals, crime victims, activists, and police embrace after the emotional premiere of *Nothing to See Here: Watts*. (Courtesy of Resnick Aspen Action Forum)

As Jefferson reminds us, the coats of our ancestors—however comfortable through long wear—will not fit the challenging weather ahead. The future demands not just new garments, but entirely new patterns for cutting the cloth.

Meeting this challenge, at the inflection of the river, is what this book is all about, and what we'll discuss further in Part Three, "Transformation." It's a task that is also now my life's work and mission, as we choose to move towards the Age of Abundance for All.

A new type of thinking is essential if mankind is to survive
and move toward higher levels. This new thinking must
be based on love and compassion for all of humanity.

—Albert Einstein

CHAPTER 4
CLIMATE CHANGE

The Sun, the hearth of affection and life,
pours burning love on the delighted earth.

—French poet Arthur Rimbaud

There's a good reason the Sun stands at the heart of humanity's oldest faiths—invoking the deity Ra in ancient Egypt, Inti among the Inca, Tonatiuh in Aztec cosmology, Surya in Hinduism, and Amaterasu in Japan's Shinto tradition. Our ancestors knew what we sometimes forget: The Sun gives *life*.

This reverence may seem an unlikely way to begin a reflection on humanity's greatest contemporary challenge—a rapidly warming climate that reveals itself in devastating storms, record heat waves, catastrophic wildfires, and the flight of millions of people from transformed landscapes around the world. Against this backdrop the ancients can still remind us to ask a fundamental question: What shall we *do* with the gift of the Sun?

Our debates about the climate crisis tend to circle around

immediate technologies and tactics. We seek cleaner fuels, more resilient buildings, electric vehicles, wind farms, and solar panels against the skyrocketing demands for electricity led by dynamic developments that include the massive AI-driven data centers we are constructing around the world. These are all urgent and vital concerns, but they orbit a deeper truth: What we call a climate crisis is, at its core, an energy crisis. And every form of energy we use—fossil, renewable, or nuclear—ultimately traces back to the Sun and solar system.

Each *hour*, our star delivers to Earth more energy than humanity consumes in an entire *year*. This ancient fusion reactor, burning ninety-three million miles away, has powered every breath of wind, every drop of rain, and the growth of every blade of grass. For billions of years, life's foundation has been nothing more—and nothing less—than this gift that enables the capture, storage, and consumption of solar energy. Coal is the compressed sunlight of vanished forests. Oil is the fossilized breath of prehistoric seas. Even uranium, the fuel of the nuclear power we must urgently expand, is part of the Sun's ancestry—a legacy forged in ancient stellar explosions and carried into the solar system long before our own star ignited 4.6 billion years ago.

**Image 4.1: Stonehenge's precise alignment with the solstice sunrise hints at a deeper purpose—as an ancient temple devoted to the sun.
(AdobeStock)**

The record of how well—or poorly—we have stewarded this cosmic inheritance fills libraries and headlines and has left our planet scarred by the legacy of strip-mined mountains, wars for oil, toppled governments, retreating glaciers, and so much more. In N'Djamena, Chad—the hometown of my friend Kallibou, the Uber driver whom you met in Chapter 2, "Abundance"—the air in 2017 was so thick with dust storms, cooking smoke, and vehicle exhaust that every breath carried particulate matter thirteen times the safe limit. That year, one in ten deaths in N'Djamena could be traced to polluted air, including the deaths of more than 7,000 children. In New Delhi, capital of India, the world's largest country, the haze of burning fuels has shortened average life expectancy by more than eight years. These are not isolated tragedies but broad symptoms of a deeper imbalance in how we capture and spend the Sun's gift.

Welcome to the Anthropocene—We Made It, We Must Fix It

Scientists have proposed a name for this imbalance: the *Anthropocene*, a new epoch that would follow the geological *Holocene* of the last 11,700 years, which began with the end of the Ice Age. The term was first popularized in 2000 by atmospheric chemist Paul Crutzen and biologist Eugene Stoermer to mark the moment in Earth's long story when our species gained the power to shape the planet's systems. For the first time, the plot is being written not by ice sheets or volcanoes but by human choices—and they are choices about energy, consumption, and ultimately the power of the Sun.

We can blame, deny, or grieve the circumstances that brought us to this existential juncture. But if the story is to end well, only we can write that ending, and we must quicken the pace of that work dramatically. Doing so requires, as we discussed in the last chapter on geopolitics, a cognitive model that is fit for this human-made epoch.

Let's turn to a helpful model that is more than a half century old but still worthy of our attention. This meta model and analytical tool was created in 1972, when a team at MIT published a paradigm-shifting report, "The Limits to Growth." You may want to clip this visual and save it in your wallet or purse, because it speaks volumes about the mission before us. Despite its age, the book captured with this graphic, "Human Perspectives," the essential challenge of our era: We tend to see planetary problems in the narrowest of timelines and proximities, when their solutions lie far beyond both.

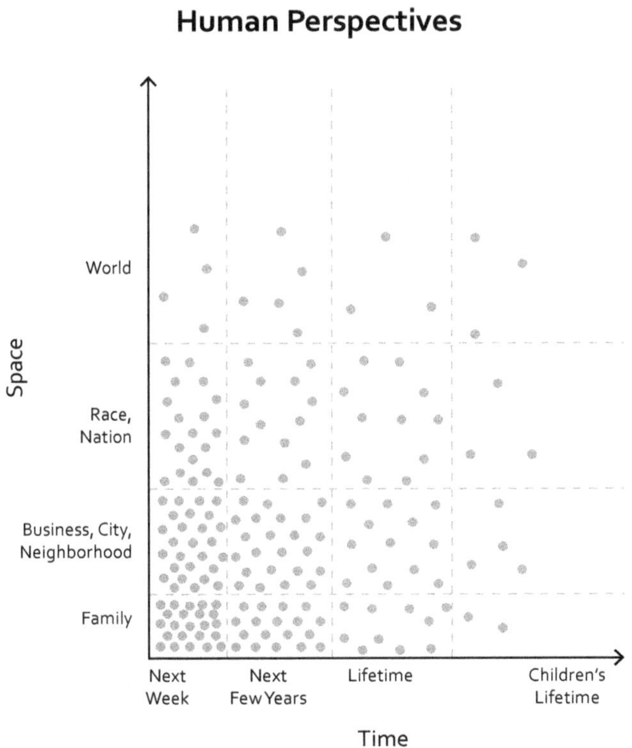

Image 4.2: *Limits to Growth*, 1972

To understand this graphic, look at its two dimensions: the horizontal axis represents Time (moving from "next week" to "the next generation" and beyond), while the vertical axis represents Space (expanding from our "family and neighborhood" to the "entire world").

As the model illustrates, the vast majority of human concern is naturally confined to the lower-left quadrant—the immediate, local "here and now" of short-term survival. However, the existential challenges of the Anthropocene require us to consciously shift our perspective to the upper-right quadrant—the global, long-term horizon where we consider the flourishing of the entire planet for our children's lifetimes and beyond.

But first and broadly, *The Limits to Growth* utilized the MIT-developed World3 computer model—one of the first major uses of computer modeling for global systems analysis—to simulate scenarios of population, industrial output, food, resources, and pollution. Its warning that population growth and resource consumption would lead to mass starvation and ultimately societal collapse if left unchecked sparked worldwide debate. The initiative, launched by the think tank known as the Club of Rome, put in motion what we today call the sustainability movement. The book was translated into thirty languages, sold more than 20 million copies worldwide, and seeded countless initiatives, including, indirectly, the Conscious Capitalism movement with which I've long been associated.

Critics today charge that the report ignited undue panic as few of its predictions panned out. As a technologist, however, I see that history differently. I see *The Limits to Growth* as a report on then-legitimate issues that were ultimately resolved, successes that now stand as a validation of the importance of technology.

The Wizardry of Technology—Then and Now

Resource depletion was offset by improved extraction methods like deep-sea drilling and fracking. Efficiency gains in industry and transport, and large-scale recycling and substitution, muted the resource threat. One of my favorite books on how pronounced this effect was is *More from Less: The Surprising Story of How We Learned to Prosper Using Fewer Resources—and What Happens Next* by my friend Andrew McAfee, co-founder and co-director of the MIT Initiative on the Digital Economy. The specter of global famine, meanwhile, was erased by the Green Revolution of high-yield crops, synthetic fertilizers, and irrigation systems, for which the architect, the late Professor Norman Borlaug of Texas A&M University, won the Nobel Peace Prize. Today's biotechnology, genetically modified crops, and precision agriculture using sensors and AI are building on that early success and doing so even as you turn the pages of this book.

"The Borlaug [approach] has become an emblem of the view that the road through humankind's environmental difficulties lies through the groves of scientifically guided productivity," wrote Charles C. Mann in his seminal 2018 book, *The Wizard and the Prophet.*

That invaluable book is effectively an update to the premises of *The Limits to Growth.* It explores the history of tension between "Wizards," who believe technological innovation can solve humanity's challenges, and "Prophets," who argue that respecting ecological limits and restraint is the only sustainable path forward. By this definition, count me among the Wizards. Importantly, Mann's book foreshadowed the exact dynamic of the March 2023 call for an AI development moratorium, which, as we discussed in Chapter 1, "Conquering Fear," was both misguided and unrealistic, for AI and its companion technologies of robotics and quantum computing are essential to steering us clear of the climate/energy crisis, as we shall see.

The central facts still with us are that the work of the MIT team, led by husband-and-wife scientists Dennis and Donella Meadows, put the issue of emissions and their environmental effects on the public agenda, and climate change remains the one threat the team identified that remains unresolved.

This returns us to the insight of the seemingly simple graphic above, which is as thoughtfully evocative today as it was when first published, for it illustrates how humanity's awareness of existential issues is generally confined to the lower left-hand quadrant. Most challenges, particularly as they are mediated through the prisms of 24/7 media algorithms and their insidious companion of increasingly shallow political discourse, lead to action only when they are immediately in front of us, or on the nearest of time horizons. In other words, in the lower left-side section of the quadrant. Ironically, the skew of our shortening attention spans toward the near periphery of concern and the shortest of contemplative timelines may be even more pronounced and perilous than was the case a half century ago.

I dwell on this model from 1972, the year I was born, because energy demand and production, once very local or regional concerns, are now animated by global dynamics. These include current AI demands but also urbanization, electric vehicles, cloud computing, the supply disruptions of war and sanctions, and growing prosperity in less-developed regions, which all make the picture more challenging. In the coming decade, we need to nearly double our energy production to meet surging demand, a feat humanity has never before achieved and one that AI should help us attain.

"The interplay of U.S., Chinese, and U.S. policies is producing ripple effects far beyond national borders, impacting supply chains, innovation, cycles and energy scenarios across the world," concludes the World Energy Council's 2025 review, a comprehensive survey of multiple energy scenarios by the London-based global energy organization.

This is not to say we should ignore local, near, and immediate threats—nor opportunities to reduce our carbon footprints. Passenger cars produce about 12 percent of global greenhouse emissions. Livestock production and consumption contribute even more, about 15 percent of global emissions, which is one of the reasons I'm a vegetarian. Keeping thermostats low in the winter and high in the summer, recycling and composting, planting more trees and more native plants are all activities in the lower left-hand quadrant that we should embrace.

The Solar Stewardship Framework

The first pillar of this transition is learning to more effectively "grow" our infrastructure through optimized photosynthesis. In the old, lower-left mindset, we mined iron ore to produce steel—a process of high-heat extraction. Today, we are seeing the rise of structural bamboo in places like Nepal and the Philippines. Bamboo is essentially a high-speed solar capacitor; it captures the Sun's energy and carbon from the atmosphere, turning it into fibers stronger than steel. We aren't just "building" anymore; we are harvesting "structural sunlight."

We also must close the loop on what I call "squandered sunlight." Every time we discard a garment from the $1.8 trillion apparel industry, we waste the energy it took to create it, energy that began as sunlight millions of years ago (as oil-based synthetics) or last year (as cotton). The transition to a "circular economy," powered by AI-driven fashion robotics that eliminate overproduction, is about energy preservation. By keeping materials in a loop, we stop the "leak" of our solar wealth. Initially developed and promoted by the Ellen MacArthur Foundation, the embrace of a circular economy by governments and businesses worldwide is an initiative I applaud.

This is explored by my friend Pablos Holman, whom you met in the last chapter. In his book *Deep Future,* Pablos explores a stunning example of the waste and destruction in the global apparel industry. For one, it's almost entirely dependent on exploitation of the cheapest labor to be found. Until the 1980s, most of the clothes sold in the U.S. were American-made. Then we moved production to China, then to Vietnam, and then to the new champion of Bangladesh, which employs 4.5 million workers in thousands of factories in conditions that would shock even Charles Dickens. Workweeks of sixty to eighty hours are common for wages that barely creep above $100 a month. "We're running out of places to go where people are willing to work for a few dollars a day," Pablos writes.

But that's only half the grim story. With crisscrossing supply chains of slow-moving components and finished garments, notoriously fickle fashion trends outpace bets on what to produce. As a result, of the 150 billion garments produced in 2024, *a third* were never worn even once. The industry, meanwhile, creates more carbon emissions than aviation and shipping combined, and it accounts for 20 percent of plastic in the ocean and a similar share of global freshwater pollution.

Until recently, automation was elusive in textile industries, where stitching requires nimble fingers—usually women's and often children's. But as Pablos points out, robots with advanced computer vision—and soon translucent fingers simulating human touch—can be trained with AI to do this sweatshop labor and we could deploy AI education tools to massively reskill those workers in far more humane jobs in healthcare, data labeling, even operation of 3D garment manufacturing equipment. Will such endeavors take large investment, creativity, and vision?

Absolutely, but we can do it. And the compounding effects of such locally focused initiatives as fast-growing bamboo are essential, along with innovation to consign outdated, brutal technologies to museums. Similarly, we need to expand our thinking and realize that renewables and fossil fuels are not in competition but need to be choreographed in tandem as a bridge to a carbon-free future. As a Texan I'm proud that my state now leads the nation in solar and wind power, surpassing California in 2024. Nationally in that same year, America passed a big milestone when we generated an estimated 17 percent of our electricity from these two sources, overtaking coal, with a 15 percent share, for the first time in history.

I'm also proud that Texans, despite perceptions to the contrary, don't see renewables as competition with conventional energy. As just one of many examples, since 2018 ExxonMobil has been partnering with Denmark's wind and solar giant Ørsted for provision of both sources of power to support its operations in what we call the oil patch. Towering wind turbines and sprawling solar farms are now as much a part of the West Texas landscape as the oil derricks and gushers made famous by James Dean in the 1956 film classic, *Giant*.

Both of these renewable energy sources, however, suffer from the sobering reality of intermittency. When the wind isn't blowing and the Sun isn't shining, our power grids need to be backed up, typically with natural gas-fueled turbines, and in some cases hydroelectric power. That's not to say solar and wind won't be increasingly important, particularly as new battery storage technologies akin to Tesla's Megapack evolve. Add nuclear power to the mix—if we can get over our cultural resistance to it (despite its 18 percent share, it's now actually a declining slice of the U.S. energy mix)—and this combination of sources can help a great deal. We'll talk more about nuclear power in a moment, but the issue remains that solar and wind technologies, while critical, do not yet reach far enough up the space-time axes.

To get our climate challenges under control, we need to think further, as the "Human Perspectives" graphic suggests, in terms of the world, our children's lifetimes, and beyond. And critically, this means that we need to acknowledge the wisdom of the ancients. We need to embrace this amazing gift of the Sun that each moment showers us with enough energy to make cheap, if not free, electricity, which will be the foundation of the Age of Abundance for All.

In short, we need to capture the cosmic and divine gift of the Sun at massive scale and learn to store it, transport it, and apply it in ways we can scarcely yet imagine. The good news is that scientists and engineers are already working to make this possible, with AI serving as the connective tissue that links each breakthrough.

The Molecular Mastery of Today's Material Science

As we climb the gradient of innovation, materials science leads the initial way. The London-based Energy Transitions Commission, a global coalition of energy producers, technology companies, financial institutions, and environmental NGOs, has called the shift in the material we make and use nothing less than a wholesale transformation of how humanity harnesses the Sun. Its recent report, "Material and Resource Requirements for the Energy Transition," forecasts that the power and efficiency of solar photovoltaics will expand more than *twentyfold* by mid-century, becoming the backbone of a global energy system powered primarily by sunlight. This surge will be fueled by panels that are increasingly efficient, less dependent on scarce inputs like silver, and supported by robust recycling of old modules to recover critical materials.

Carrying and balancing this new flow of electricity will demand transmission grids that triple in size and density, new types of ubiquitous batteries to store energy, and electrolyzers to produce green hydrogen fuel. Virtually carbon free and with the technology to do this imminent, hydrogen will soon be drawn from seawater. Hydrogen, in fact, is the most promising fuel to reduce the carbon footprint of long-haul jets. ThyssenKrupp, the German steelmaker, is piloting new techniques to use hydrogen in steel production, technology that could lead to a 7 percent decrease in global carbon emissions. Notably, ThyssenKrupp is doing this with use of the Siemens Industrial Copilot, which is just one in a suite of generative AI-driven, industrial copilots built on Microsoft's Azure OpenAI services.

Each of these technologies—most available today but constrained by political and social inertia—depend on copper, lithium, nickel, and graphite, posing new challenges. I'm not suggesting this will be easy. The rare earth elements such as yttrium, critical for superconductors, or neodymium and praseodymium, essential for the permanent magnets in wind turbines and EVs, exist in few places and are prey to familiar geopolitical hoarding. The search is on for battery components such as cobalt on the ocean floor, igniting concerns for environmental destruction of marine life and fisheries. But even these challenges represent new opportunities for innovation, substitution, and circularity.

AI, accelerating computer power, and ultimately quantum computing are emerging not just as mere side stories but as enablers of this materials transition. AI models already optimize solar panel performance, guide precision mining, and stabilize power grids in real time. Even without the unimaginable speed quantum computing will bring, my friend Pablos argues that the materials underlying computer science are undergoing their own unheralded revolution. Silicon is yesterday's medium. Infinitely faster chips made from graphene, or even diamonds, are just around the corner, he predicts. Moore's Law, the famous premise that chip speeds double roughly every two years, is fast being eclipsed by what Holman calls "Jensen's Law," in honor of NVIDIA founder and CEO Jensen Huang, that will enable acceleration chip speeds far beyond today's fastest. "Every day, I see new algorithms, new chip architectures, new ways of making computers faster," Holman wrote, "a hundred times faster, a thousand times faster, sometimes a million times faster."

The Quantum Leap Promised by Quantum Computing

Beyond these mind-bending advances in materials lies quantum computing, a radically new form of computation that harnesses the laws of quantum mechanics—superposition, entanglement, and interference. Quantum computers will process information in ways no classical computer can. Instead of the familiar digital bits of 0s and 1s, quantum bits (qubits) can exist in multiple states at once, allowing certain problems to be solved exponentially faster than on today's supercomputers.

In the spring of 2024, I attended the annual TED conference in Vancouver and witnessed a fascinating talk by Hartmut Neven, founder of Google Quantum AI. With his signature sunglasses atop his head as a permanent headband, wearing disheveled denim straight out of a cyberpunk novel, he has the swagger of someone more expert in the physics of a Harley than the nature of the universe—though I suspect he's expert in both. He explained the company's newly unveiled quantum chip, dubbed *Willow*, which Google claimed in December of 2024 will outperform any supercomputer on Earth (note that it was bested since then by other quantum-chip efforts multiple times over.) Though still a research prototype, Willow could be accessible via the cloud before the decade is out, Neven suggested. In one benchmark, *Willow* performed a computation in under five minutes that would take one of today's fastest supercomputers 10^{25} years, that's *ten septillion* years, to complete. "This mind-boggling number exceeds known timescales in physics and vastly exceeds the age of the universe," Neven wrote in a blog post. Actually, it is *725 trillion times* more than 13.8 billion years, the hypothesized age of the universe. Even more striking, he argued that this feat validates a hypothesis outlined by physicist David Deutsch in his 1997 book, *The Fabric of Reality*, that quantum computers derive their massive computational power by performing calculations simultaneously across multiple parallel universes. The strands of this technology transform the ideas in the "Human Perspectives" graphic into three dimensions. In Deutsch's reasoning, which I'm fascinated by, each universe contributes a different computational path before the results converge back into our observable world. This is a radical idea we'll explore further in Part III, "Transformation."

**Image 4.3: Computing in parallel universes.
Google's Hartmut Neven hypothesizes
at TED 2024 in Vancouver.
(Courtesy of TED Conferences, LLC)**

Quantum computing promises to accelerate breakthroughs in materials science, from discovering new superconductors to designing catalysts that make hydrogen production cheaper and more efficient. Together, these technologies act as accelerants, representing intelligence applied to matter and energy, collapsing decades of trial and error into months of simulation and design.

As the Energy Transitions Commission concludes: "Solar efficiencies will keep rising steadily, reducing the materials and land needed for every gigawatt installed." With AI and quantum computing driving that curve upward, a constellation of emergent technologies is within humanity's grasp.

The promise of the Sun's power is so fundamental that Pablos has quipped that we should add it to the base of Abraham Maslow's famous hierarchy of needs. "If you solve energy, you could solve a lot of other problems for free," he writes. "With abundant, clean, and cheap energy, you can desalinate seawater to make freshwater for everyone.

You can power the extraordinarily energy intensive atmospheric carbon capture devices that people are inventing. You can grow high-quality safe, nutritious food for everyone."

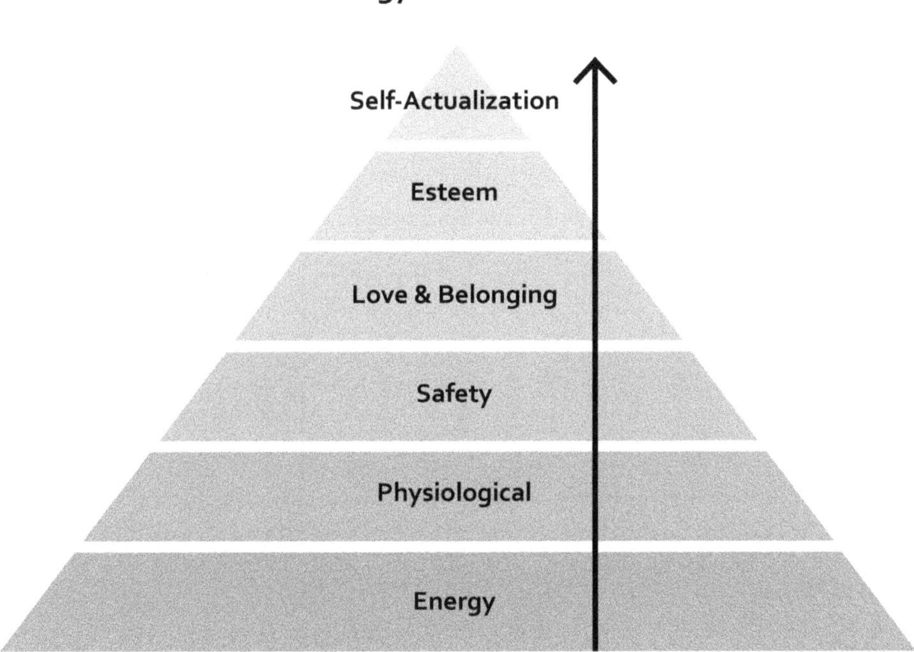

Image 4.4: Maslow's iconic hierarchy updated to reflect Pablos Holman's innovation.

Let the Turbines, Centrifuges, and Cooling Fans Spin

While these applications represent critical bridges to truly abundant energy, America finds itself falling behind in the one technology capable of delivering it at scale. Nuclear power, or fission, is the proverbial elephant in the room of energy discussion. It is a sector once led by the United States. America launched the world's first commercially-scaled nuclear power plant in 1957, but our dominance peaked in the 1980s and now the sector is led by China, France, Russia, and Korea.

A fraught topic, nuclear power has been stalked by decades of fear and emotion. It's a bit ironic that the name of the 1979 film that stirred anti-nuclear passions was *The China Syndrome.* It imagined a meltdown so severe it burned through the Earth "all the way to China."

In the movie, heroic nuclear engineer Jack Godell, played by Jack Lemmon, unsuccessfully tries to warn the world before he is gunned down by a SWAT team. His sacrifice isn't meaningless, however. Intrepid journalist Kimberly Wells, played by Jane Fonda, manages to broadcast some of Godell's warning before she too is thwarted, and we never really find out if the earth survived. Yes, dramatic accidents like Three Mile Island, where the reactor core partially melted just three weeks after the movie premiered highlighted genuine dangers and led to a host of design improvements and new safety regulations. But the containment structure held, no one was injured or killed, and today Microsoft is trying to put together a deal to buy the power from the damaged plant's adjacent reactor, which was shut down for other reasons and could be restarted in 2028. Microsoft has also signed a deal with the ambitious nuclear startup Helion, which aims to launch the world's first nuclear fusion reactor by 2028.

As the World Energy Council points out in the report mentioned earlier, the danger of death or injury from coal-fired energy is 1,000 times *greater* than that of nuclear power, wind, and solar power; the risk is *100 times* greater from gas-driven power turbines. As Pablos puts it: "We conflated nuclear reactors with nuclear bombs and outlawed the wrong one."

Now, of course, we face a real "China Syndrome" of a different nature. The U.S. is still the #1 producer of nuclear energy, generating about a third of the world's total that meets roughly 18 percent of America's demand. But we won't be leading for long as we operate with the world's oldest nuclear infrastructure, and the number of plants is down from 100 to 93, with most of those units close to a half century old. The most recent facility came online in Georgia just in 2023, but only after three *decades* of delays and staggering cost overruns that have spooked the industry. Meanwhile, in the past decade, China has added 34 nuclear power plants, for a total of 58 reactors. The country has 30 more under construction, another 36 planned, and more than 150 proposed. In China, they come in on time and on budget. By contrast, the U.S. has *just one plant* under construction and no more than ten on the drawing boards. A recent summary from the International Energy Agency declared, "In five years, China will overtake the U.S. and Europe and will be the number one nuclear power in the world."

Economist and Nobel laureate Paul Krugman addressed this issue in a column in late 2025: "Does this mean that the U.S. is losing the race with China for global leadership?" he wrote, answering: "No, I think that race is essentially over."

I certainly hope Krugman is wrong and I'm not quite as pessimistic. But ponder this assessment against the two graphs below:

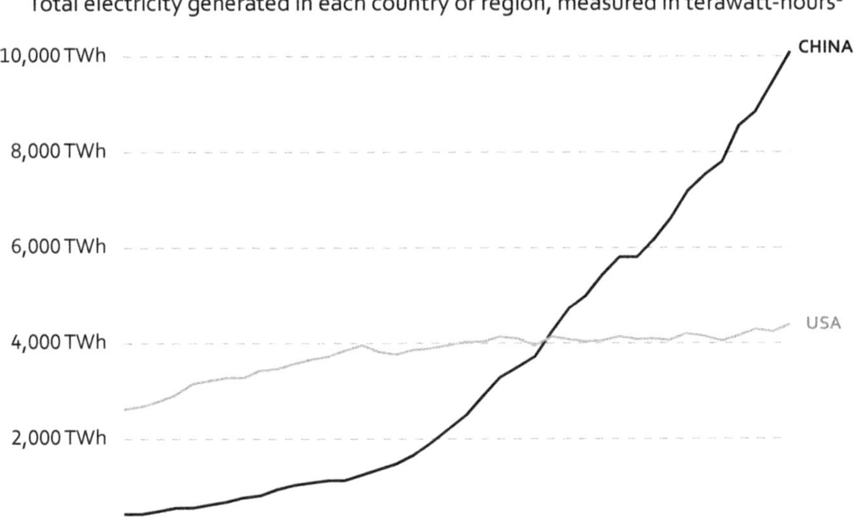

Data source: Ember (2025); Energy Institute - Statistical Review of World Energy (2025)

1. **Watt-hour** A watt-hour is the energy one watt of power delivers for one hour. Since one watt equals one joule per second, a watt-hour equals 3600 joules of energy.

Metric prefixes are used for multiples of the unit, usually:
- kilowatt-hours (kWh), or a thousand watt-hours;
- Megawatt-hours (MWh), or a million watt-hours;
- Gigawatt-hours (GWh), or a billion watt-hours;
- Terawatt-hours (TWh), or a trillion watt-hours.

Image 4.5: Balance of Power (Our World in Data)

U.S. Nuclear Power Capacity

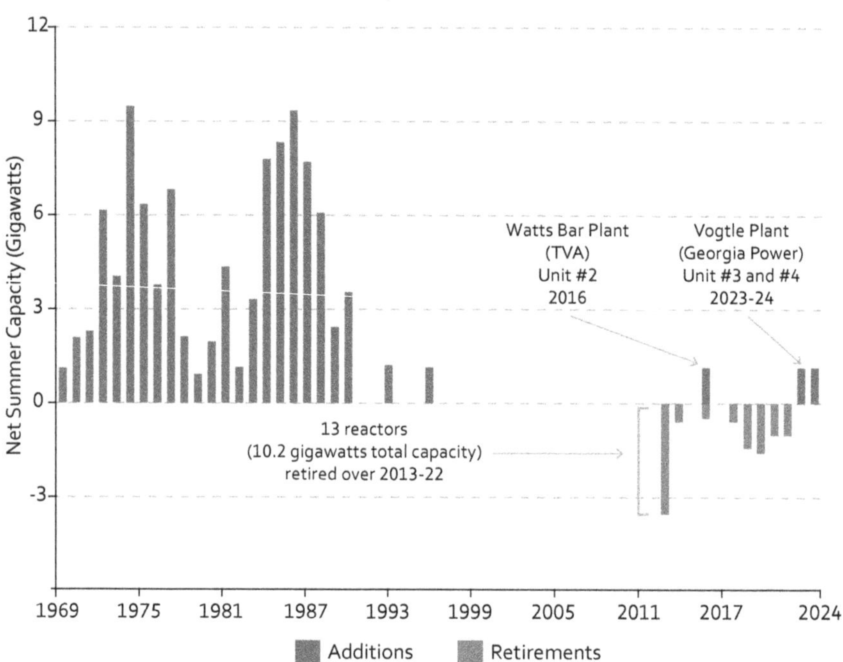

Image 4.6: Additions and Retirements, 1969-2024 (The National Center for Energy Analytics)

Here's another way to think of it. America's total nuclear capacity is 97 gigawatts, or GW. That's enough to power 75 to 100 million American homes for a year. It's also equivalent to just *18 seconds* of sunlight. China's capacity, meanwhile, is 55.3 GW, or 10 seconds of sunlight on a clear day. We can add to our anxieties about China with the fact that its massive and rapidly growing reactor fleet will surpass our generating capacity, or we can get moving ourselves.

And we might. A comeback appears to be in the making. These could come with investments in advanced reactor designs and small modular reactors (SMRs). To be clear, there are two overlapping technologies involved in the looming renaissance. One category is the "conventional" SMRs that, like their larger, workhorse predecessors, are mostly water cooled. These are smaller, factory-made, and passively safe reactors, akin to suits off the rack rather than those made by bespoke tailoring. Standardized and rapidly deployable, these technologies promise rapid reduction of the costs of new atomic power plants—if we can produce them at scale.

The other new technology, broadly, is so-called fourth generation or "Gen IV," which includes some versions of SMRs that use gas cooling, molten salt coolants, and sometimes fuels that replace uranium rods with spherical uranium "pebbles" that make runaway chain reactions and catastrophic meltdown an even more remote possibility than in conventional designs.

So far, only one SMR reactor and one pebble-based Gen IV version have been put into commercial operation. The first SMR, in Russia, is actually a floating variety, its design copied from the nuclear ice-breakers that have been in use since the first was launched in 1957. It was assembled atop a barge and has been generating power since 2020. China, meanwhile, switched on the world's first pebble-based Gen IV reactor in 2023. As soon as 2026 China will launch the first reactor that is genuinely *both* Gen IV and an SMR.

As of late 2025, the Trump administration was pushing a host of initiatives with the goal to quadruple nuclear power generation in the next twenty-five years. These chiefly seek to accelerate U.S. nuclear power generation with slashed regulation, faster licensing, and tax breaks. What's missing is serious cash.

The private sector has stepped into the gap. Gen IV nuclear start-ups are drawing interest, not just from VC and infrastructure funds, but also from Big Tech, the catalyst being obvious—the skyrocketing power demands of AI and data centers. Alphabet has partnered with Kairos Power LLC, an advanced nuclear reactor developer, to deploy 500 megawatts by 2030. Amazon partnered with X Energy and Dominion Energy on reactor projects, and Meta Platforms is hunting for nuclear power as well. An S&P Global Market Intelligence analysis found that private equity transaction value in the advanced nuclear sector reached a record $783.3 million in 2024, that's 13x the 2023 total.

One of the most interesting studies of America's nuclear logjam comes from the non-partisan National Center for Energy Analytics, founded in 2024 in my home town of Austin. That study, "A Strategy for Financing the Nuclear Future," suggests we need an entirely new project financing model for the new and emerging technologies—a proposal that resonates with me. In short, the old model, of nuclear plants financed by power utilities themselves, is essentially dead. A new model, the research argues, should come from the large infrastructure private equity firms—the Blackstones or KKRs of the financial world—who need to step up to the plate. "Currently, some $1 trillion of infrastructure-related private equity capital is available that could be used to fund greenfield U.S. nuclear projects," the report concludes. The study further argues that the equity powerhouses alone have the requisite experience in the overlapping domains of energy, infrastructure, project delivery, and risk management—along with ample capital resources—to pull America out of the nuclear doldrums. I'm hopeful.

More Efficient Use of the Sun Is Not *Just* About Electricity

We are also upgrading the "Solar Engines" of our food supply. Building on Norman Borlaug's Green Revolution, projects like the C4 Rice initiative are using AI to re-engineer how plants handle sunlight. By upgrading the plant's internal software to be 50 percent more efficient at converting light into calories, we move from the fear of famine toward true nutritional abundance.

As you may recall from your high school biology class, most plants use a "C3" version of photosynthesis, in which the first product formed contains three carbon atoms. An enzyme called Rubisco does the carbon-grabbing, but in heat or dryness it often grabs oxygen by mistake, causing wasteful "photorespiration" that burns energy and slows growth. Rice is a C3 plant, as are many staple crops, and this works well in cooler, wetter conditions. In hotter, brighter, drier climates, however, some plants—like corn and sugarcane—evolved a "C4" system that first makes a four-carbon compound and concentrates around Rubisco. This avoids the oxygen mistake, runs more efficiently, and delivers higher yields. Rice missed that evolutionary train, leaving it less efficient under heat and drought. So the idea behind "C4 rice" is to retrofit rice with this CO_2-concentrating machinery so it behaves more like corn.

Oxford's C4 Rice Project is progressing but was still in the research-and-prototype stage as this book was going to press. Its scientists are now using AI and computational biology to design, integrate, and test C4 traits faster and more precisely, pinpointing gene targets and modeling leaf metabolism before trialing them in the field. If successful, the implications are enormous: up to 50 percent higher yields with improved water and nitrogen efficiency, dramatically reducing the strain of rice farming on rivers and aquifers. For the more than three billion people worldwide who depend on rice as their primary food, this could mean greater food security under climate stress, while simultaneously increasing carbon capture and freeing land for other uses in Asia and Africa.

In essence, C4 rice is not simply an agricultural advance; it is another expression of the same principle that runs through solar panels, batteries, and material science: finding ever more ingenious ways to capture the Sun's energy. Whether through silicon cells on rooftops or engineered leaves in a rice paddy, humanity's task is the same, to turn sunlight into abundance and move us further up the gradient toward meeting challenges.

Beyond C4 rice, a whole suite of technologies known as "living Sun-harvesters" is emerging at the intersection of biology and technology. So-called biomimicry technologies such as algae fuels and bio-photovoltaics use microalgae to turn sunlight into fuels or even electricity. Artificial photosynthesis and artificial leaves being developed at MIT split water and convert CO_2 into usable energy carriers. Agrivoltaics pair crops with solar panels to maximize land use and conserve water, and biohybrid solar cells or "bionic leaves" combine engineered bacteria with catalysts to make liquid fuels from sunlight. Together, these advances point to a future where we harvest solar energy not just through silicon panels and reactors, but also through living systems—leaves, microbes, and algae—designed to capture the Sun's power with nature's own efficiency.

The Ultimate Gamechangers—Nuclear Fusion and Space-Based Solar Power

This game-changing suite of urgently needed technologies we've explored is not the ultimate embrace of the Sun's power. To reach the far horizon of "Human Perspectives"—the upper right of our "Human Perspectives" cognitive gradient, where solutions endure across generations—we must look to two technologies that mirror the Sun itself, or place us in its unbroken path.

The first is nuclear fusion, the very process that powers the Sun and stars. Unlike today's fission reactors, which split atoms, fusion merges light nuclei to release staggering amounts of energy with little waste. Just a gallon of seawater contains enough deuterium—an isotope of hydrogen critical to fusion—to yield the energy equivalent of 300 gallons of gasoline. Several recent milestones underscore how close we are to mastering fusion technology.

In 2023, the U.S. Lawrence Livermore National Laboratory twice achieved "ignition," a net energy gain where fusion output exceeded the energy delivered by lasers. In Oxfordshire, England, the Joint European Torus, or JET, laboratory set a record of 69 megajoules—enough to heat the typical American home for about twenty hours—in a five-second burst. That marked the most powerful controlled fusion reaction on Earth to date. And here, AI is not just an enabler but a necessity. Reinforcement learning systems are already being used to stabilize plasma in reactors, preventing disruptive events and extending reaction times. As we've discussed, with private-sector investment certain to surge fusion is moving from science fiction toward commercial reality. As mentioned, Microsoft has even signed a deal for power with the ambitious nuclear startup Helion, which aims to launch the world's first nuclear fusion plant by 2028. It's a moonshot kind of technology, yet it promises a nearly inexhaustible, carbon-free energy source, literally mimicking the Sun on Earth.

A second emerging energy technology is space-based solar power, or SBSP. Its advocates envision capturing sunlight in orbit where it shines without interruption and beaming it wirelessly to Earth. The idea was first championed by Peter Glaser in the 1960s and has long been dismissed as fanciful. But in recent years it has leapt back into the realm of the possible. In 2023, Caltech's Space Solar Power Demonstrator proved that lightweight deployable structures could unfurl in orbit. The project tested dozens of solar cell types under space conditions, and, most dramatically, it wirelessly transmitted power across space. In late 2025, Japan succeeded with the first full-scale demonstration of beaming solar energy down to Earth. The implications are staggering: Power stations in geostationary orbit could deliver clean electricity day and night, unaffected by clouds, seasons, or geography, turning the whole Earth into a single grid lit by the Sun's unbroken flame.

Together, fusion and space-based solar stand as the twin apexes of our technological climb. One recreates the Sun's fire here on Earth. The other extends our reach into orbit where the Sun never sets. If achieved at scale, they would not only meet humanity's energy demands but also redefine our relationship to the cosmos itself, completing the arc begun when ancient peoples first raised their eyes to the sky and named the Sun a god with many names.

For millennia, humanity has worshiped the Sun as giver of life, building myths and rituals around its rising and setting. Today, we stand at the threshold of capturing that gift not just in fields and forests, but in silicon wafers, engineered leaves, and in reactors that mimic its fire and satellites that orbit in its endless light. Fusion and space-based solar are more than technologies; they are symbols of a civilizational choice. They ask whether we will remain bound to scarcity, conflict, and short horizons, or whether we will embrace abundance—energy so vast it can lift billions from poverty, cool a heating planet, and power the dreams of generations yet unborn.

When we succeed, we will have fulfilled the oldest human intuition: that the Sun is not only to be revered but to be shared, and its light converted into food, fuel, and flourishing. Seizing the gift, scaling it beyond imagination, and making it not just for survival, but for the Age Abundance for All is the task of our century.

The Civilizational Choice: The Great Synthesis

Together, fusion and SBSP represent the apex of our technological climb toward energy abundance. We must recognize that this progress was born from a century-long "contestation" between two fundamental human archetypes: the Wizards and the Prophets.

The Wizards have delivered the means for our survival—from the Green Revolution that fed billions to the efficiency gains of the internet and the AI-driven material science we see today. Yet the Prophets provided the conscience. While their darkest warnings often failed to manifest, like Paul Ehrlrich's best-selling 1968 book *The Population Bomb* that predicted millions would starve by 1980, their victories were essential. It was the Prophets who fought for the Endangered Species Act, who cleaned our air and water, and who championed the early moral case for solar power long before it was profitable.

The Age of Abundance for All is the moment when these two forces realize their conflict was a false choice. The Prophet and the Wizard are finally joining forces. The Prophet's prayer for a healed Earth is being answered by the Wizard's mastery of the Sun's fire. We are finally moving past the era of Squandered Sunlight—those long decades where we starved in the midst of plenty because we lacked the tools to catch the light.

When we look at the upper-right quadrant of our potential, we see that the "Energy Crisis" was always an "Efficiency Crisis." By applying AI to the fundamental building blocks of matter—whether by re-engineering a rice leaf to drink more sunlight or capturing the fire of the stars in a fusion reactor—we are moving from a world of extractive scarcity to one of radiant abundance.

We are no longer merely passengers on a finite planet; we are becoming the architects of a system where technology and nature coexist in harmony. By uniting the Prophet's wisdom with the Wizard's ambition, we are ushering in the Age of Abundance for All, and it is no longer a distant dream—it is our current path of innovation, where both now stand shoulder to shoulder to ensure the Sun is never squandered again.

*I love to think of nature as an unlimited
broadcasting station, through which God
speaks to us every hour, if we will only tune in.*

**—Scientist and inventor
George Washington Carver**

It is the sea itself who fashions the boats,
choosing those which function
and destroying others.

—French philosopher Émile Chartier

On July 16, 1945, the sky lit up over the New Mexico desert, and a mushroom cloud reached 7.5 miles into the sky. The flash of the so-called Trinity bomb was visible 200 miles away. As the Manhattan Project's director J. Robert Oppenheimer recalled years later, his immediate thought was of a line from Hindu scripture's *Bhagavad Gita*: "Now I am become Death, the destroyer of worlds." All of us are familiar with the history of that moment.

Since that day, hundreds of books have been written on this dawn of the nuclear age, and more than fifty major films have sought to capture it, including the 2023 blockbuster *Oppenheimer* by Christopher Nolan. We know where this has led: Today, nuclear arsenals in nine countries, bristle with more than 12,000 warheads, including those unveiled in 2025, such as China's "full nuclear triad" of land, air, and sea-based missiles, and America's new-and-improved B 61-13 nuclear bomb for use against harder and large-area military targets. Shockingly, it would take just 2 percent of *today's* arsenal to destroy all our major cities across the world and send humanity and all the planet's animals into a global nuclear winter for a decade. The skies would darken, it would be freezing cold, and *billions* would die. Yet, we are producing more warheads today for the purpose of ensuring Mutually Assured Destruction, a military posture ironically represented by the widely used acronym MAD.

This doctrine of MAD—justified a half-century ago as a psychological gambit that allowed adversaries to deter one another—has become as outdated as rotary dial handsets and switchboards in a world of ubiquitous smartphones. As artificial intelligence and autonomous weapons systems compress decision-making cycles from minutes to milliseconds, we find ourselves without a doctrine of war and peace fit for an always-on, algorithmically mediated era.

Compounding the growing threat, the world's last remaining guardrail—the New START treaty first signed in 2010 to limit strategic U.S. and Russian nuclear warheads—expired on February 5, 2026. While the existing nuclear arsenals already posed an unacceptable threat to humanity, without the New START treaty the risk of nuclear use will likely increase with a heightened nuclear arms race. Its expiration marks the first time in fifty-plus years that there are no binding limits on the strategic arsenals of the two largest nuclear powers.

Meanwhile, the role of *fear,* that primal force we began to explore at the outset, remains inadequately understood and is seldom if ever confronted. Imagine an alien who had just landed on Earth. They would quickly zero in on the reality that, with this level of fusion-and-fission-powered weaponry, humanity is, indeed, MAD and likely marvel at what an interesting acronym MAD is to emphasize the disease we collectively need to cure and transcend.

Little captures this disparity between today's reality and a mindset locked in a vanished age as a May 2025 remark by America's Energy Secretary Chris Wright, announcing the completion of America's newest nuclear bomb: "Modernizing America's nuclear stockpile is essential to delivering President Trump's peace through strength agenda."

Our need to contend with today's prospect of accidental annihilation by a machine error, a faulty sensor, or a software glitch was better framed by UN Secretary-General António Guterres' remark in 2022: "Today, humanity is just one misunderstanding, one miscalculation away from nuclear annihilation."

The list of countries flirting with the idea of joining this extraordinarily dangerous nuclear club is growing—led most notoriously by Iran, whose nuclear program continues to alarm rivals and would-be international regulators. Others include Saudi Arabia, whose crown prince has said the kingdom would pursue nuclear weapons if Iran did. Turkey and South Korea have begun serious debate about nuclear options. Poland has publicly urged the United States to deploy nuclear weapons on Polish territory, while Germany has begun exploring a broader European "nuclear umbrella" debate amid transatlantic uncertainty. Ukraine, which surrendered its Soviet-era arsenal in 1994, and Japan have both seen renewed public and elite discussion of nuclear options, including nuclear sharing or hosting arrangements. Even in the Nordic countries of Norway, Sweden, and Finland, analysts and commentators have begun asking whether a "Nordic nuke" should be discussed—all these discussions signal just how far the taboo has shifted.

Russia, meanwhile, not to be outdone, has repeatedly taken to rattling its amped-up nuclear sabers, including a nuclear-tipped nuclear missile that is also nuclear *powered,* giving it unlimited range and loitering time. That development was stunning when unveiled to the world personally by Putin in March 2018. It makes the renewal of America's space race that began with the Soviet launch of Sputnik in 1957 all that more understandable—if myopic. Amid its ongoing war with Ukraine, Russia also updated its nuclear doctrine in late 2024 to allow use of nuclear weapons in response to "non-nuclear aggression backed by a nuclear power."

Essentially, the world is doubling down on MAD's grim principle that no country will launch a first strike, as doing so guarantees its own annihilation by the retaliation of the enemy. The system is stable only if three conditions are met:

1. **Certainty:** The retaliatory strike is certain to occur.
2. **Capability:** Both sides must maintain the capacity to retaliate (a robust "second-strike" capability).
3. **Time:** Decision-makers (humans) must have enough time to assess the situation, verify an attack, and consciously decide whether to retaliate.

When Forgetting Becomes Dangerous

All this logic, if ever logical, is now obsolete. However insane this six-decade trajectory, the history of MAD is worth a quick review.

Before you started reading this book, did you know that only 2 percent of today's nuclear weapons would destroy all major cities and kill billions of us? I didn't either when I began researching and writing. Yet no films have been made, and only a handful of mostly academic books have been written about what is, to my mind, the moment of equivalent importance to that split second in the New Mexico Desert—this one happened at a press conference in London's Caxton Hall.

On July 9, 1955, philosopher Bertrand Russell and a group of scientists led by Albert Einstein (who had died after signing three months before) released the "Russell-Einstein Manifesto." Widely regarded today as the dawn of the nuclear abolition movement it is a document whose words resonate profoundly more than seven decades later. The *Manifesto* explored the implications of the then-recent test of a hydrogen bomb on the Bikini Atoll in Micronesia's Marshall Islands—the prototype of a device which remains the most powerful bomb the U.S. has ever exploded.

Dubbed "Castle Bravo," the bomb unleashed more than 1,000 times the destructive force as the fission bombs that destroyed Nagasaki and Hiroshima. Its explosive yield was badly miscalculated, and the Bikini blast also contaminated nearby islanders, some of the 28,000 U.S. service members overseeing the test, and the unfortunate crew of a Japanese fishing boat, the *Lucky Dragon.* Even decades later, Bikini

Atoll remains uninhabitable, and cancer cases in nearby islands are widespread. In 2022, then-President David Kabua of the Marshall Islands unsuccessfully appealed for retroactive compensation and justice at a White House summit of Pacific island leaders with President Joe Biden.

It was this very test that prompted Russell and Einstein to issue their historic warning. "We have to learn to think in a new way," Russell, Einstein, and nine other scientists wrote, referencing the Bikini Atoll test of 1954, which demonstrated their assessment "that nuclear bombs can gradually spread destruction over a very much wider area than had been supposed."

The authors of the *Manifesto* continued: "We have to learn to ask ourselves, not what steps can be taken to give military victory to whatever group we prefer, for there no longer are such steps; the question we have to ask ourselves is: what steps can be taken to prevent a military contest of which the issue must be disastrous to all parties?"

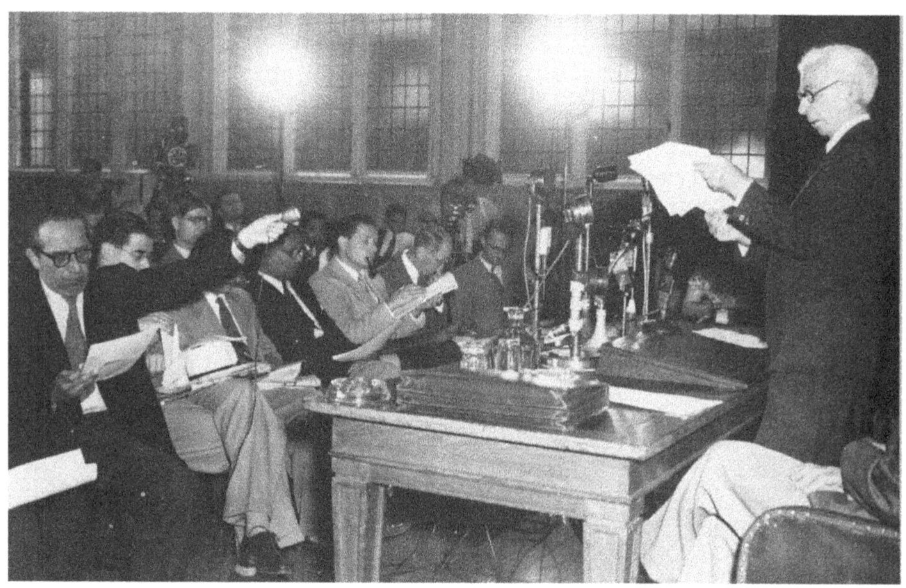

**Image 5.1: Bertrand Russell reads the
Russell-Einstein manifesto on July 9, 1955,
(Photo: Pugwash Britain)**

While now obscure, the message of the Russell-Einstein Manifesto was heard in important circles. Its most immediate result was the 1957 Pugwash Conference in Nova Scotia, a gathering of scientists that continues as an annually convening think tank to this day. Together, the *Manifesto* and Pugwash really set the disarmament movement we know today in motion, resulting in the Partial Test Ban Treaty in 1963 and the Nuclear Non-Proliferation Treaty of 1968.

More broadly, the imperative to cope with the arms race that has ensued in the years since is best understood as two distinct tracks—and we need to explore both.

The Architecture of Deterrence

This first track hardened into the infrastructure of deterrence: redundant weapons systems, silos in Montana (offensively positioned for the shortest range to Moscow via the North Pole), strategic bombers in Texas (defensively positioned for the maximum over-the-pole distance from Soviet strike range), research to weaponize space, alliances, doctrines of "flexible response," and an evolving arms-control regime. Out of the ethos of weapons manufacture emerged the governmental pursuit of equilibrium, what we came to call MAD. Key milestones included:

- **SALT I (1972)**—The first Strategic Arms Limitation Treaty capped U.S. and Soviet strategic ballistic missile launchers, including both ICBMs and submarine-launched ballistic missiles (SLBMs).
- **SALT II (1979, signed but never ratified)**—The accord still has been observed for years by both superpowers.
- **START I (1991)**—The Strategic Arms Reduction Treaty mandated actual deep reductions in deployed nuclear arsenals, a first. The agreement limited both sides to 1,600 delivery vehicles

carrying no more than 6,000 warheads.

- **START II (1993, signed but never entered into force)**—The treaty would have banned Multiple Independently Targeted Reentry Vehicles (MIRVs) on ICBMs and reduced strategic warheads to 3,000–3,500 for each side. The U.S. ratified it in 1996, and Russia ratified it in 2000 with conditions, but Russia withdrew in June 2002 after the U.S. withdrew from the ABM Treaty.
- **New START (2010, expired February 5, 2026)**—The treaty limited both sides to 1,550 deployed strategic nuclear warheads and 700 deployed delivery systems. Russia suspended participation in 2023; the treaty expired on February 5, 2026, ending over fifty years of continuous nuclear arms control between the superpowers.

Together, these treaties defined the rhythm of the Cold War arms race—restrained but never stopped. This pursuit of treaties seeking balanced lethality did reduce the number of warheads maintained by the two main nuclear rivals. Yet improvements in accuracy, survivability, and delivery systems mean that today's smaller arsenals are in many ways more militarily effective and strategically destabilizing than ever. Above, we explored how MAD's doctrine emerged in the era of rotary phones, but a more useful analogy here is the cell phone. At the signing of Salt II, at Motorola's launch of the world's first cell phone in 1979, the demo weighed two and a half pounds and had a talk time of twenty minutes. Today's smartphone has an average talk time of six hours, and weighs about 175 grams, less than a third of a pound.

Analogously, across the same time frame we have enhanced the lethality of nuclear weapons through reductions in size, densification of mass, and miniaturization of components. Ironically, this architecture has made nuclear war more thinkable. Our challenge is to use AI not to accelerate this fast pass to the graveyard, of which Guterres warns us. Rather, AI must center a new architecture of innovative collaboration.

The Architecture of Global Disarmament

Meanwhile, running in parallel with deterrence has been the moral, social, and intellectual movement pushing not for nuclear equilibrium, but for abolition. This tandem track includes the birth of the 1960s peace movements, protests, think tanks, NGOs, and countless initiatives. Just a few include America's Union of Concerned Scientists, the Stockholm International Peace Research Institute, Switzerland's International Campaign to Abolish Nuclear Weapons, and many more. Milestones include:

- **The 1982 Nuclear Freeze Campaign:** The massive U.S. grassroots movement pressed for a halt to new nuclear weapons.
- **The International Physicians for the Prevention of Nuclear War (founded 1980, Nobel Peace Prize 1985):** These medical experts warned of the humanitarian consequences of nuclear war.
- **The Treaty on the Prohibition of Nuclear Weapons (2017, entered into force 2021):** The first legally binding international agreement to comprehensively ban nuclear weapons was largely symbolic. Only non-nuclear powers signed the accord; all nuclear powers boycotted the negotiations producing it; sadly, the result is akin to vegetarians agreeing to meat-free Fridays.

This idealistic but still incomplete architecture is less about hardened silos and more about the soft power of moral suasion, civic activism, and global norms that stigmatize the bomb.

Somehow, this dual track architecture has kept the nuclear bombs from flying. But neither architecture is complete without fully confronting that foundation of our destructive impulses that we explored at the beginning of this book: fear. To confront this primal driver behind both deterrence *and* disarmament we must acknowledge how deeply rooted and how dangerous it is.

I recognize that I've rough-sketched a great deal of complex history into this summary, and I'll come in a moment to what a new doctrine beyond fear to replace MAD could look like. But I also don't mean to dismiss the complexities of the Cold War or diminish the stakes and the strategic thinking and leadership that ultimately led to the collapse of the Berlin Wall. This ranged from George Kennan—the diplomat who authored the original 1940s doctrine of "containment" to challenge Soviet expansionism without sparking a war—to President Reagan, whose 1982 proposal to transcend MAD with the Strategic Defense Initiative (SDI) was nicknamed "Star Wars." SDI was a technological attempt to eliminate the threat, thereby making the weapons irrelevant. While SDI as envisioned was to languish, its spirit of foreseeing a "planetary sensorium"— a concept mentioned in Chapter 3, "Geopolitics", that we'll further discuss below—may guide us yet.

And we should acknowledge scholar and former Secretary of State Henry Kissinger, whose work to keep the nuclear genie bottled up continued until he finished his final book, *Genesis—Artificial Intelligence, Hope, and the Human Spirit*, just weeks before his death at age 100. I credit all these important contributions to our avoidance of calamity, so far.

Genesis was Kissinger's swan song and love letter to humanity, where he acknowledged right before he passed that fear itself is our primary danger. "Neither blind faith nor unjustified fear can form the basis of an effective strategy," he wrote, along with co-authors Eric Schmidt, the former CEO of Google, and Craig Mundie, the former head of research and strategy at Microsoft, "One needs self-doubt to have knowledge, but self-confidence to act," this trio of authors continue. "Indeed, in the age of AI, this is all the more urgent. We must try to understand the challenges that AI will present even as we lack the prior exposure or the essential experience to guarantee the accuracy of our comprehension."

Mundie also advised his friend and longtime collaborator, *New York Times* columnist Thomas Friedman, on a very compelling column that argued China and the United States will inevitably overcome their fear of one another, as AI brings us together. "Our conversations [with

Mundie] over the past 20 years have led us to this shared message to anti-China hawks in Washington and anti-America hawks in Beijing: 'If you think your two countries, the world's dominant A.I. superpowers, can afford to be at each other's throats—given the transformative reach of A.I. and the trust that will be required to trade A.I.-infused goods— you are the delusional ones,'" Friedman wrote in September 2025.

Equally important, I don't want to neglect the courage, insight, and inspiration of the nuclear abolitionists, from physicist and Nobel Peace Prize winner Linus Pauling to Hiroshima survivor Setsuko Thurlow, still active at this writing as a disarmament campaigner at age ninety-four.

A remarkable history of this activism, and a call for a new global ethos for disarmament, is Indian scholar Sundeep Waslekar's book, *A World Without War—The History, Politics, and Resolution*, which resonates with our current tensions of nationalism and polarization. First published in 2022, *A World Without War* is particularly notable for its exploration of what Waslekar calls "weaponized hyper-nationalism" that seeks domination through, again—that basic impulse—fear.

Waslekar is a co-signatory of the "Normandy Manifesto for World Peace" of 2019, a declaration that he issued with four Nobel Peace Prize laureates—Mohamed ElBaradei, Leymah Gbowee, Denis Mukwege, and Jody Williams—that calls for a new global social contract rooted in cooperation and human dignity. "Domination instigates fear and violence and needs implements," Waslekar warns. "The greater the ambition for dominance, the deadlier the instruments that are required." My respect for these strategists, leaders, and activists is immense. But I offer this caveat: Reduced to fundamentals, these two tracks, deterrence versus abolition, combine as a kind of Möbius strip—seemingly two-sided but ultimately a single band of fear-driven progression. The whole is greater than the proverbial sum of both parts; we have created an edifice that merges both tracks—a global architecture of fear. But, today, we need an ethical, global architecture for collaborative innovation—one that transcends both the obsession of the world's militaries with technological advantage and the limits of a disarmament movement that has run its admirable course.

A Global Architecture of Fear

To create this ethical, global architecture for collaborative innovation, we need to look beyond both MAD and abolition—this unified architecture built on fear—to approaches that steer us away from catastrophe. We need guidance not from purely rational and material thinking, but from ancient wisdom. Here I find inspiration in the *Tao Te Ching*, the 2,500-year old Chinese philosophy that underlies both Confucianism, Buddhism, and has repeating patterns throughout Christianity, Judaism, and Islam. As we've seen, humanity is approaching a crucial turning point—a bend in the river of consciousness—that demands a higher perspective, one where our task is not to choose sides in this debate. Instead, we must heed what Russell and Einstein urged us to do decades ago: "We have to learn to think in a new way."

As we discussed in Chapter 1, "Conquering Fear," every great binary eventually yields to unity—matter and energy, wave and particle, Self and other—to which we can now add deterrence and abolition. This is a principle of both modern science and of ancient wisdom, and we see it most clearly in the verses of the *Tao Te Ching*, the foundational text of Taoism attributed to the philosopher Lao Tzu (老子), who lived in the sixth century BCE. The translation by Wayne W. Dyer ✌ is like a workbook, offering daily exercises to go along with each verse. The translation by Stephen Mitchell is modern and pure, and Mitchell's voice on the Audible version is soothing and contemplative. ✌

There are many translations and interpretations of this text, but I have found the most actionable is that of the late psychologist Wayne W. Dyer, a scholar of Lao Tzu and the *Tao Te Ching*, whose many books include *Change Your Thoughts—Change Your Life*. In Dyer's reading of Verse 1 of the *Tao Te Ching*, he draws on its lessons to note that Western thought typically seeks to resolve paradoxes, while Eastern and non-dualist philosophy seeks to accept "paradoxical thinking" and move above and beyond—as we must do to escape the paradox of our nuclear Möbius strip. "Paradoxical thinking is embedded in Eastern concepts, such as yin and yang or the feminine and masculine, and where things are comfortably described as both this, and that," Dyer writes. "Whereas we in the West, by contrast, tend to view opposites as incompatible concepts that contradict each other."

I encourage everyone to read the *Tao Te Ching* (sometimes translated as Dao, as it is pronounced in English) in its entirety. From this Verse 1 introduction, let's move to Verse 31, which is the response—two millennia earlier—to the call of Russell and Einstein. It reads in part:

> Weapons are the tools of violence. All decent men detest them. Therefore, followers of the *Tao* never use them. Arms serve evil. They are the tools of those who oppose wise rule. Use them only as a last resort. For peace and quiet are dearest to the decent man's heart, and to him, even a victory has no cause for rejoicing. One with a will to kill will never prevail upon the world. It is a good sign when man's higher nature comes forward, a bad sign when his lower nature comes forward. With the slaughter of multitudes, we have grief and sorrow. Every victory when you win a war, you celebrate by mourning.

In a sense, the world's current insistence on developing autonomous, AI-driven deterrence and execution systems can be summarized in the *Tao* as the *Way of the Knife*: precise, competitive, and sharp—exactly the danger we must avoid. Rather, the *Tao's* perspective urges us to seek the *Way of the Balance*—not the "equilibrium" of MAD but a harmony of purpose. Let's pray that both the U.S. and China follow the *Way of the Tao* when it becomes most clear that they must.

Our AI Future Is Ours to Choose

There is no denying that AI holds the potential to exacerbate the risk of nuclear annihilation. This is not because AI itself is inherently evil—it is not. It is because humanity can use tools that are unprecedented in their power to either create massive destruction or blissful abundance. Technology has always been an enabler for humanity. When we discovered our original technology of fire, we primarily used it to cook our foods, which increased our IQ, and to warm our homes. But we also knew it burned hot and would lead to excruciating death if applied to human or animal flesh. It was used in that way too, to be sure. But never forget that the core nature of humanity is loving and must be rediscovered as such.

The risks of AI, including the exacerbation of nuclear and other threats, is a subject that we'll explore more deeply in Chapter 11, "Risks." But imagine for a moment the counterfactual. What if nuclear fission had debuted as a technology for power generation and we had never invented the bomb? What if the Manhattan Project had been judged too costly or we had added that $35 billion (in today's dollars) to the $160 billion to rebuild post-war Europe with the Marshall Plan?

In one sense, such imaginings are a college coffeehouse debate like asking *What if England had won the American Revolution? What if the Bolsheviks had never come to power in Russia? What if Lee Harvey Oswald had missed on that fateful day in 1963?* But serious historians *have* considered many what-ifs, including Richard Rhodes, the Pulitzer Prize-winning author of *The Making of the Atomic Bomb.* ❦ Who asks such questions as *What if Nazi Germany had gotten the bomb first? What if the U.S. had pursued openness and an early international bargain instead of secrecy and monopoly? What if the postwar control plans had been adopted before the arms race hardened?*

Among the leading scientists who imagined a different path was physicist Joseph Rotblat, who left the Manhattan Project in protest in 1944, once it became clear the Nazis did not have a viable nuclear program. ❦ He went on to become the first secretary-general of Pugwash, running the movement and organization from its founding until 1973 and earning the Nobel Peace Prize. Even after the dropping of the Hiroshima and Nagasaki bombs, opinions differed in the inner circles of the world's nuclear scientists. In Rhodes' recounting, Oppenheimer famously squared off with his physicist colleague Edward Teller, insisting that no more bombs should be made and, "We should give Los Alamos back to the Indians,"[*] Oppenheimer insisted. Teller went on to become known as the "father of the hydrogen bomb," first tested in 1952 and 700 times more powerful than the bomb dropped on Hiroshima.

It's a complicated history of course, but a different path, as envisioned by Rotblat and Oppenheimer, might have been possible. Imagine if the Manhattan Project had pivoted with all its collective brilliance to a global endeavor of electrical production: Coal burning might have been phased out by 1970. Electricity would be cheap and abundant. Climate change might still be a challenge, but it would be a far lighter burden. Rivers like the Colorado (which I rafted in the summer of 2025 within the 277 miles of the Grand Canyon) wouldn't be drying up as we would be desalinating freshwater on the coast and piping it where needed.

Human Agency: War Is a Choice, Not Destiny

When it comes to taming the nuclear threat in the era of AI, we are not without thoughtful ideas to create a future very different from the one toward which we are currently and dangerously heading, for the powerful tide of the Superfecta of AI, robotics, quantum computing, and BCI—Chartier's "sea" that fashions our boats—is forcing us to discard the old, leaky vessels of deterrence.

What must emerge is a third path—neither deterrence nor immediate disarmament, but a new architecture of strategic innovation and collaboration, a way of using AI to reduce the likelihood of catastrophe by design, rather than gambling on fear or faith.

Among the most intriguing proposals is offered in the paper "Generative AI and WMD Nonproliferation: A Practical Primer for Policymakers and Diplomats,"[*] produced in 2024 by the James Martin Center for Nonproliferation Studies. The report rejects the premise that AI's value lies in faster reaction or automated response. Instead, it argues that AI should be deliberately used to *slow escalation rather than speed it up*. It suggests fusing early-warning data while explicitly quantifying uncertainty. It envisions a global architecture to detect anomalies in military activity and deception that generate false alarms. It recommends verification checkpoints rather than automated action—buying time for deliberation rather than compressing it. This primer further proposes using AI to strengthen verification and transparency by monitoring compliance, flagging deviations, and reinforcing confidence-building measures.

The compelling report, by security analyst Natasha E. Bajema, also draws a bright line against any use of AI in launch authority or escalation-driving decisions. In effect, the report sketches a new architecture in which AI is not a trigger, but a brake—a modern institutional expression of the *Tao's* vision and a call for restraint—designed to preserve the human pause on which survival has always depended.

Among the cluster of promising concepts that could buy us time is the framework of Minimum Deterrence (MD), championed by strategic thinkers and scientists like Alexander Glaser and Zia Mian. MD is certainly a better acronym than MAD, and suggests we could use a doctor to cure our disease. This approach proposes that nuclear-armed states commit to radical stockpile reductions—potentially to a "low-baseline" of just a few hundred warheads—maintaining only what is necessary as a deterrent of last resort against existential threats. While not the immediate abolition sought by many, MD represents a decisive step toward global stability. This model is made increasingly realistic through advancements in AI and remote sensing, which can solve the verification challenges that stalled past proposals. Much as Dr. Bajema has advocated for a global architecture to detect WMD-related anomalies, a tech-enabled verification regime could subject the world's remaining fissile material to constant, high-fidelity scrutiny, ensuring security without the need for massive arsenals permanently on high-alert.

A broader and more ambitious framework for situating these nuclear-specific proposals appears in *Children of a Modest Star—Planetary Thinking for an Age of Crises*, by Jonathan S. Blake and Nils Gilman, which we also discussed in the last chapter. Crucially, they present a much evolved articulation of Reagan's intuition that led to the vision of SDI, what they call a "planetary sensorium," a global nervous system of satellites, monitoring stations, sensors, and other emerging technologies allowing us to perceive the planet as a whole.

I reviewed the book shortly after it was published in 2024, in the essay, "How the Ethos That Led to the First Thanksgiving Should Guide Us in the New Age of Discovery." *Children of a Modest Star* is a highly original work in which Blake and Gilman begin from a sobering diagnosis: Many of our inherited global institutions—the United Nations, the World Bank, the International Monetary Fund, even the annual climate summits—remain trapped in short-term sovereignty games, ill-suited to governing risks that unfold at planetary scale and across generations. Rather than calling for a new UN bureaucracy or world state, Blake and Gilman sketch a more plural architecture—overlapping regimes, polycentric governance, and new constitutional principles—that encode duties to future generations and the biosphere itself.

Crucially, Blake and Gilman treat emerging technologies like AI, synthetic biology, and geoengineering not as side issues, but as stress tests for planetary governance. Their concept of "planetary realism" accepts power as it exists, while insisting that survival now depends on stronger knowledge-driven institutions, longer time horizons, and our willingness to design innovative governance systems that would favor collaboration and investing in an infrastructure built upon harmony. When read alongside the ideas of Bajema, Glaser, and Mian, Blake and Gilman's work suggests that AI's most important role may not be to optimize power, but to help humanity learn—at last—how to govern our own strength.

If we remain stuck in the competitive, zero-sum model, we will use our new tools—the Superfecta—to accelerate the arms race, driving us toward a chaotic, machine-error-induced extinction. But there is clearly another way.

As author and scholar Waslekar argues in his book *A World Without War* mentioned above, the issue is broadly a matter of choice: "Countries wage wars because they want to," he writes. "Geography, history, economy, and other such factors can provide legitimacy once the choice is made. But pretext is not the same as cause. Choice—mostly a calculated and conscious one—is the sole cause of war."

This is precisely how I regard AI. We must choose. As we evolve our collective consciousness, and we must, we will make the right choices so that AI becomes the means of humanity's flourishing. Humanity has never had technology this powerful that was accelerating so exponentially. In the wise words of prophets from Voltaire to Uncle Ben in the *Spider-Man* movies, "With great power comes great responsibility."

The Test of Our Generation: The Final Exam for Humanity

Let's close with the hope that the *Tao Te Ching* will again be China's most celebrated ancient spiritual guide. China today, of course, is not moving in the spirit of the *Tao* any more than other players in the new nuclear arms race are. If anything, its mobilization toward the full nuclear triad discussed above is clearly focused on automation, early warning, and compressed timelines.

With China and the U.S. being the world's AI superpowers, and those creating the most advanced nuclear arsenals, my prayer for humanity is that we walk *the Way,* the path of the *Tao's* moral guidance. This is the test of our generation, preparation for God's final exam for humanity: Do we use abundance-level technology to multiply scarcity-level fear, or do we do the opposite?

We often must turn back, remembering what has been lost, to advance toward what can be gained going forward. The Age of Abundance for All beckons and passing God's final exam for humanity is crucial to our survival on this planet. Discovering our spiritual cores and the nature of our actual reality, which is deeply rooted in love, will be what I explore most in Part III, "Transformation."

The atomic bomb has altered profoundly the nature
of the world as we know it, and the human race
consequently finds itself in a new habitat
to which it must adapt its thinking.

—Nelson Mandela

If we teach today's students as we taught yesterday's, we rob them of tomorrow.

—John Dewey, philosopher and educational reformer

I'm going to begin our discussion of education with an assertion that you may see as pure heresy: There is no "education crisis." At least not in the sense we often use that term in reaction to a long list of distinct maladies. Of course, there are likely one or more crises ongoing in your children's school, or at your university alma mater, and crisis is certainly an apt description of the situation in much of the developing world where nearly 300 million children get no schooling at all. But our focus on these serious, plural crises should not distract us from the singular "meta-crisis" that is at the root of them all: the collision between what I call linearalism, a term I invented in a series I wrote in mid-2023, "Why the 'Education Crisis' is a Global Malady (and How AI Can Fix

It),"🕊 and the non-linear, exponential world of AI.

This might seem like an abstraction, even a case that would get me shouted down at many a school board meeting or faculty retreat, by both the political right *and* the left. But before you angrily slam this book shut, let's mull over just a few of the many things going on.

The Failure of Linearalism

My home state of Texas was riven in 2025 by wrenching debate over vouchers, the growing practice in numerous states to fund private schools with so-called Empowerment Scholarship Accounts, or ESAs (meaning public money). Whatever your view on vouchers, tune into E. Ray Moore, the nation's leading ESA advocate. As he told *Harper's Magazine* in the well-reported and thoughtfully-reasoned 2025 article, "The Homemade Scholar—The New Frontier in American Education," 🕊 "Right now, the number of children outside the public school system is around 15 percent. If we can get that percentage to 25 percent, the [public] system would start to implode."

Moore's aspiration might soon be fulfilled. Texas recently became the fifteenth state to launch a voucher program; as of this writing, at least ten more states are considering one. As I was beginning this chapter, headlines everywhere were trumpeting the latest National Assessment of Educational Progress, a biannual report dubbed the "nation's report card." It delivered the grim news that twelfth graders were producing their worst reading and math scores in two decades 🕊, just the tip of an iceberg in a report recounting much more—all equally disturbing. Blame smart phones, pandemic overhang, burned-out teachers, exhausted parents, or the hunger with which an estimated 5 million elementary schoolers in America start their day.

Meanwhile, a shocking World Bank report 🕊 just a few months earlier revealed that in sub-Saharan Africa four out of five children can't read a simple text, war has shuttered 14,000 schools, 100 million illiterate African children have never been to *any* school, and the need for 15 million new teachers is urgent—a figure more than three times the entire number of K-12 teachers in the United States.

Compounding the challenge, let your mind go back to our opening chapter on fear. The preponderance of debate around all these issues refracts through the prism of politically-manipulated, algorithmically-driven, polarizing fear. 🕊

The Real Crisis—And the New Concept of Sphericism

In the face of imploding public schools, collapsing scores, an urgent demand to deploy a cadre of new teachers in Africa equal in number to the population of the Netherlands, hungry children in America, and spreading fear, you are reasonably asking: If this is not a crisis, what is?

To ignore this is heresy for sure. I'm not dismissive of very real problems. My heart breaks over much of what I see, and over that which I'm sharing here. We have many crises in the way we acquire knowledge, diffuse it, use it, and pass it. It is just that as a pretty serious believer in the utility of data, and having founded and run six data-driven tech companies, I see and use the word *crisis* advisedly and always seek to discern what lies below—in the meta.

The *meta-crisis* is not in education. Rather, it is at the interface between these very real challenges and the very limited scope of the solutions in our existing educational nomenclature and store of ideas. This interface operates like a broken API, tech jargon for application programming interface: The inputs are real, the outputs are needed, but the protocol in between is corrupted.

To fix this corrupted protocol, we must first name the two competing operating systems currently battling for our future—linearalism and my other coined term, *sphericism*—representing the learning ecology we need. We must transition from an obsolete nineteenth-century mindset to a twenty-first-century reality:

Linearalism (The Factory Model)
- **The Model:** A factory assembly line where students move at the same speed in age-batched

cohorts.
- **The Goal:** Compliance, standardization, and the production of "safe" cogs for a predictable industrial economy.
- **The Metric:** Standardized testing and a linear A to F grading scale.
- **The Constraint:** Resource scarcity—limited teachers, fixed teaching hours, and unchanging textbooks.

Sphericism (The AI Learning Web)
- **The Model:** A non-linear, multidimensional "learning web" that reflects the intricate networks of the brain.
- **The Goal:** Cultivating both "Prophets" *and* "Wizards"—curiosity-driven problem solvers who can evolve and collaborate in a rapidly changing world.
- **The Metric:** A "living map" that illustrates relationships, proximity, and shared skills instead of rigid rankings.
- **The Constraint:** Radical abundance—availability of AI tutors, global peer networks, and the limitless broadcasting potential of nature.

Without this shift in our API, every attempt to fix our schools is simply an attempt to build a better version of a broken machine. We can see this failure clearly when we look at the global landscape.

Both linearalism and sphericism represent a broad category of problems that we will explore shortly. But for my purposes here, it is linearalism—a close cousin to the "scarcity mindset" we explored in Chapter 2, "Abundance"—that obscures our understanding of education. More specifically, it distorts how we've understood learning since the mid-nineteenth century, when we began building institutions around what has now become an outdated educational paradigm.

Understanding "Half-Lives of Knowledge" and Root Causes

This is the real crisis with three dimensions, all of which we'll unpack:

- First, we've handicapped our ability to confront the fast-changing dynamics of learning with education systems little changed since the middle of the nineteenth century.
- Second, we too often ignore the collapse of what information scholars call the "half-life of knowledge." This is the reality that today the utility of knowledge expires almost as quickly as it is produced; the shelf life of expertise is measured in months, and no longer in decades.
- And three, these issues of the changing learning dynamics and the collapsing half-life of knowledge converge into the sheer inertia of linearalism, this one-step-at-a-time process that has become our habit.

Academic readers will recognize my argument in intellectual debates about twentieth century *Taylorism*—Frederick Taylor's scientific management approach to workplace efficiency—and *Fordism*, the idea drawn from Henry Ford's revolutionary assembly line that transformed manufacturing worldwide. But the essential point is that we reflexively run twenty-first-century challenges—particularly education challenges—through our suite of nineteenth- and twentieth-century solutions, hoping that somehow success will emerge.

It won't.

The work of Harvard psychologist and my friend, Ellen Langer, helps explain why this is so stubbornly true. In her landmark 1989 book *Mindfulness*, Langer demonstrated that our schools don't merely fail to teach critical thinking; they actively train students *away* from it. This is an essential insight that should be at the heart of our education debate and discussion.

When a child is taught that a fact simply *is*, rather than that it *could be* or *might be under certain conditions*, the institution is training that child to stop noticing. Every standardized test, every fixed rubric, every single right answer on a multiple choice exam is a small lesson in what Langer calls mind*less*ness—the condition of operating on automatic pilot, applying yesterday's categories to tomorrow's problems without ever questioning whether they fit. Echoing John Dewey's quotation at the top of this chapter, Langer diagnoses the school system as a mindlessness factory, and the industrial model we built it on virtually guaranteed this outcome.

In her brilliant 2023 book *The Mindful Body*, Langer extended this argument into medicine, showing that mindless acceptance of fixed categories—a diagnosis delivered as absolute verdict rather than probabilistic possibility—can itself become a self-fulfilling prophecy, with measurable physiological consequences. The educational parallel is direct and urgent: the student who is taught to receive knowledge rather than question it, to accept expertise rather than test it, arrives in a world of collapsing half-lives of knowledge utterly unprepared for the one skill that the world demands above all others—the capacity to actively notice when what they know no longer applies. Langer's antidote is what she calls conditional thinking: the habit of framing knowledge as "this *could* be true" rather than "this *is* true," a small linguistic shift that her research shows dramatically improves learning, memory, creativity, and adaptability. It is, in essence, the cognitive architecture that sphericism requires.

Since I wrote the first draft of this chapter, I've gone to Bhutan on a family trip and I explore the lessons I learned on that trip in "New Models for Humanity" in Part Three, "Transformation." Model 1 in that section is on Bhutan, "The Education of a King, and the Remaking of a Nation's Measure." But, back to my point, that then and now, if everyone is having an "education crisis," perhaps there's something going on that can't be addressed with vouchers, more or fewer tutors, more rigorous testing, 15 million more teachers in Africa, and more of the hundreds of angry school board meetings that are now a feature of American life. It's like trying to talk about California wildfires, European heat waves, deepening drought in Ethiopia, floods in Texas, and hurricanes in Florida without considering—or even realizing—the global effects of climate change.

Let's break this down further.

As we move toward the Age of Abundance for All, AI will be the central tool in the suite of technologies enabling what I have called a "new ecology of learning." And this would certainly include addressing the urgent situation in Africa for which the hopes to mobilize 15 million new teachers into conventional school systems is a castle in the air. In fact, AI is already transforming learning in Africa as we discussed in Chapter 2, "Abundance," where I shared the successes of Africa's first multilingual Small Language Models, or SLMs, designed to support African languages such as isiZulu, Yoruba, Hausa, Swahili, and isiXhosa.

As I framed the dilemma two years ago, what I saw then—and what is even more apparent now—is that linearalism is not isolated to Texas, or even the United States. It is *everywhere*. The real culprit that we are avoiding is our institutional habit of processing exponential, nonlinear challenges with outdated, step-by-step tools. Not only are the dynamics of learning fast changing, the shelf life of knowledge grows shorter by the day.

The term the "half-life of knowledge" is one I've borrowed from physics, where it refers to the time it takes for half of a radioactive substance to decay. We measure radioactive isotopes this way because they don't decay at a steady rate. Nuclear decay is not like dieting to lose a pound a week, thereby trimming four pounds off your torso in a month. Instead, a constant percentage of remaining atoms decay in each time period. Hence, carbon-14 with a half-life of 5,730 years, will decay by half in that much time. It will decay another 25 percent in 11,460 years, and another 12.5 percent by 17,190 years.

The first to apply this borrowed formula to education was Samuel S. Dubin, a professor of psychology at Pennsylvania State University, who in 1972 made the case for continuing education—a case similar to the one I made in my first book, *The Entrepreneur's Essentials*, where I argued similarly in my chapter "The Always Be Learning Life."

To bring Dubin's notion up to date, imagine you were a freshman engineering student in 1950. AutoCAD and LIDAR were still decades off, transistors had barely left the laboratory, and when my grandfather,

James Mann Hurt, taught analytical mathematics at the University of Texas at Austin in the age of slide rules, the methods he taught would lead for the next three decades.

The half-life of engineering knowledge back then was thirty years. Why, you could finish off your senior year, get a job with the Texas Department of Transportation, and not crack a technical book until the 1970s or beyond. Not anymore. Imagine you're an aspiring freshman engineering student in 2026, just plopped down at UT's Ernest Cockrell Jr. Hall. Guess what? Depending on the field, the half-life of most engineering knowledge, particularly in software and electronics, is just one to three years. In computer engineering it's probably even less. Much of what you learn in that freshman year will be obsolete when you don your cap and gown in 2030.

My point here is that we no longer live in a linear world and probably never did, metaphysically speaking. We live on a quantum, networked planet where knowledge doubles in days, not decades. And our schools, still marching to the old rhythm, are left hopelessly out of sync, like Ford's assembly line: K-12, freshman to senior, bachelor's to master's to PhD. Professors' careers follow a tripartite caste system of assistant to associate to full, and everyone's performance gets ranked on the same five-note, linear scale, in use since the 1890s: A, B, C, D, F.

Tripping Over the Same Bad Math as Rome's Aqueduct Builders

The situation we've gotten ourselves into is analogous to the "water crisis" of the Roman Empire, about which I wrote in the spring of 2025, "The Digital Aqueduct: What Rome's Hydraulic Failures Teach Us About Data."

The Romans were famous for (among other things) their vast networks of aqueducts. These were engineering marvels that carried water for miles to sustain cities, baths, and agriculture. But as masterful as their infrastructure was, the Romans lacked a scientific understanding of fluid dynamics. In brief, the Romans' inadequate math assumed a linear relationship between aqueduct size and flow: double the size of the pipe, double the flow of water. Not quite. The Hagen-Poiseuille equation describes how flow increases with the fourth power of the radius, so doubling the radius increases the flow rate by a factor of sixteen.

Without access to such knowledge when things failed—when water stagnated, overflowed, or never arrived—physical collapse was often wrongly blamed. This was because no Roman understood how the system worked as a whole. The curse of linearalism stalked Roman water management, just as it stalks our dysfunctional education systems today. The Romans even contrived ceremonies to appease the "water gods" without apparent result. Many of the proposals we hear today to fix our supposed education crisis are of comparable value—as long as we continue to ignore the limitations of linearalism.

A New Model of Sphericism

Clearly, wrenching our two-dimensional schools and schooling models from the throes of linearalism and into the three dimensions that are imperative is a formidable task, but we can't get to the Age of Abundance for All without doing so, which is why I have a conceptual model to at least get us started. Again, the pace of change is outstripping the evolution of language, so I want to expand on the new term I coined above: sphericism. Like linearalism, it doesn't exist in any dictionary, but I believe this is the design philosophy, or mindset, to help us begin to think in the critical third dimension. With sphericism I suggest a model of knowledge, skills, and institutions as multi-dimensional, non-hierarchical spheres rather than linear ladders. This concept emphasizes relationships, adjacency, and continuous re-navigation over rank, grade, or sequence.

We already see hints of sphericism emerging in the technologies that most closely approximate human cognition. One of these technologies is the "knowledge graph"—the "brain" of data governance that, along with data catalogs, was core to the company data.world, that I co-founded in 2016 and led as CEO until it was acquired by ServiceNow in 2025. In fact, I wrote a series about that technology back in 2022, "Dawn in the

Anthropocene—The Evolution of Data," , where I made this analogy. The knowledge graph is similar in many ways to the Large Language Models, or LLMs, like ChatGPT, Claude, or Gemini, with which we are all now familiar. Both work in ways akin to the neural networks of the brain and operate not by climbing steps but by moving across multidimensional spaces of meaning.

Technically, knowledge graphs and LLMs navigate through adjacency, proximity, and resonance. Said differently, they learn. They re-center themselves in response to shifting contexts rather than advancing up a fixed hierarchy. Just as an LLM predicts the next word by exploring a neighborhood of relationships, or a knowledge graph connects nodes through webs of association, the brain lights up constellations of connections when recalling or creating. Sphericism, then, is not an abstraction but a name for the way intelligence—artificial and biological—already works: non-linear, relational, and spherical.

A learning ecology, as opposed to mere education, might look like this in sphericism:

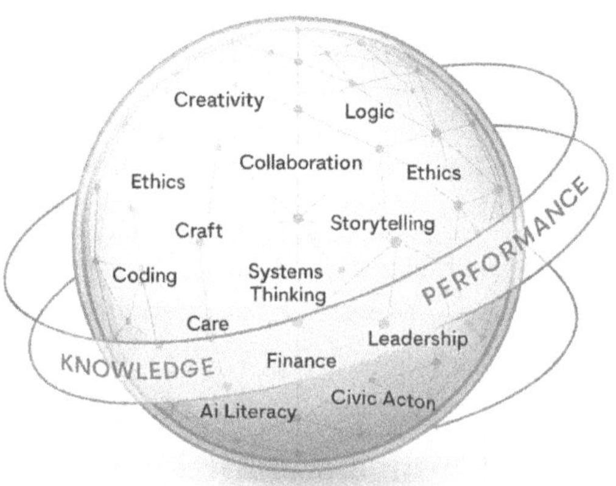

Image 6.1: The Learning Ecology of Sphericism

We're already seeing examples of sphericism emerging in education. The pandemic gave us a glimpse of what is possible when necessity forces change. Overnight, universities went virtual. Some schools, like the University of Texas at Austin, with its Department of Computer Science of more than 6,000 students, could roll out all the requisite technology, tools, and training quickly and migrate to remote learning. But small institutions like Huston-Tillotson, a historically Black university in Austin with a total enrollment of about 1,000, survived not because of administrators or faculty. It managed to quickly make the transition because students seized the bull by the horns by stepping in to train their professors on Zoom and the supporting Learning Management System, or LMS. The students and professors at Huston-Tillotson co-created a new learning model.

I've witnessed this brilliance with my own teenage child Yuzu. Three years ago, he was the first to bring OpenAI's ChatGPT to the attention of his educators at Eanes (one of the top-rated school districts in Texas). Sadly, they huddled on the potential of the tool and then quickly banned it completely from being used on campus. Undeterred, Yuzu, then fourteen, would finish his conventional homework on the bus before getting home, and then dive into this new technology. He trailblazed across this new learning frontier, developing a self-directed, personal "AI Hackathon" and explored the video game Minecraft's "modding" with AI. Already a self-taught coder, he mastered so-called "vibe coding" and invented a host of new online games. But he also didn't disappear into some online abyss or suffer the "cognitive de-skilling" that some AI pessimists warn will be the result of using this new technology. On the contrary. He explored the world in new, non-linear ways and now has explored a passion in cultural anthropology, which may be his major when he goes to university.

Another amazing AI and education pioneer is Wharton School Professor Ethan Mollick, who responded to the launch of ChatGPT not by banning it but by ambitiously redesigning his classes around it. His students now generate business plans, simulations, and startup concepts at a scale that would have been impossible before ChatGPT became the ubiquitous presence used by at least 800 million people every week.

What once took months now takes days, and the result is not diminished rigor but expanded imagination.

Similarly, Sal Khan, the founder of the Khan Academy, whose videos already educate millions, has launched "Khanmigo," an AI-powered tutor and teaching assistant. For the first time in history, a personalized tutor is within reach for every student—not just the wealthy few. I've admired Khan over the years through the annual TED Conference, where he recently became head curator, and I share the view he made in his presentation there, "How AI could save (not destroy) education."

If we listen carefully, the signal is clear: The future of learning will not be bound by classrooms, diplomas, or standardized tests. We must not think of education through the rigid models of the past—as a grid, a scale, or an axis. Instead, we must embrace sphericism—seeing education as a sphere rather than a ladder, a form with no fixed top or bottom, no single entry point, and no hierarchy of subject over subject or teacher over student, one that is equally accessible from any direction and turns at the speed of discovery, matching the exponential curve of change.

The challenge is whether our institutions can keep up, or whether they will remain prisoners of linearalism while the world around them transforms exponentially.

I realize too that we must think in new ways about how to develop and nurture what education scholars call the "hidden curriculum," the socialization, the common values, the shared identity, and the stake in society that is also a critical part of education. Perhaps this will come as a commitment to public service. Perhaps this will be the strengthening of the virtual communities that young people find and create on their own in the world of gaming. Perhaps we reconceptualize our schools and school buildings as community skill centers; not elementary, middle, and high schools, but as places where all the skills and competencies on the *sphere* of a new learning ecology will be available to all. Teachers will become learning guides.

One innovative newcomer, the Alpha School founded in Austin, is already attempting to do exactly that, transforming teachers into guides and hoping to make AI core to its pedagogy rolls out its model nationwide. Alpha School may be the boldest experiment yet to reimagine the architecture of learning. It has rapidly grown from a modest home-based cohort into a planned network of schools serving hundreds of students in major cities nationwide. At its core is a radical restructuring of the school day: just two hours of AI-driven academic work, tailored precisely to each student's level, followed by long stretches devoted to practical skills, collaborative projects, and what the school calls "masterpieces"—deep explorations aligned with a student's passions.

The teachers-turned-guides, freed from the demands of grading and lesson planning, focus instead on coaching students as they design food trucks, build video games, or even develop business ventures. It's an early success that may well inspire similar models and, more importantly, new thinking about new models.

As I suggested at the outset, there is no education "crisis" that is distinct from the "meta-crisis." We're at the beginning of a transformation the likes of which humanity has never seen. There is much to imagine, and much to do, as we resolve the paradoxes that lead us to think about tomorrow in yesterday's terms.

Of course, our reach for sphericism as the model of a new learning ecology must be paired with the conquering of fear and the embrace of love. We must produce abundance such that no child in America or anywhere will come to school hungry, and those 300 million children in Africa and elsewhere will all have access to the world's best learning tools. We'll need new models of governance and national sovereignty. We'll need to wake up and rid the world of nuclear and other weapons of mass destruction. And we need to do a lot more than that as we continue to conquer fear in the chapters still ahead.

These are the tributary streams of a new consciousness as we approach humanity's bend in the river toward a universal vision of the coming Age of Abundance for All.

The test of truth in life is not whether we can remember what we learned in school, but whether we are prepared for change.

**—German education pioneer
Andreas Schleicher, often
referred to as
"the world's schoolmaster"**

Modern medicine would do better to measure its success by the degree to which it eliminates the need for its own services.

—Philosopher, author, and social critic Ivan Illich

On our path to the Age of Abundance for All, healthcare presents a paradox: Unlike other sectors where abundance is unambiguously good, medical abundance can harm—making restraint, not excess, the measure of quality care. We are moving from a system of "sick" care, which is predicated on the presumed value of volume—more tests, more procedures, and more hospital beds filled—to a true system of "health" care. In this new era, the ultimate luxury isn't a better hospital; it's never needing one. Evocative of ancient insight, we are pivoting toward the value of health, where success is defined by the "superior doctor" who prevents the fire rather than the "inferior doctor" who merely battles the flames.

So far, most of what we've explored has been about *more*: more food, more clean water, more computing power, more energy from the Sun, and more opportunities to make life meaningful through AI, which leads the Superfecta. Healthcare in the Age of Abundance for All inverts this paradigm: fewer visits to the doctor, fewer if any prescriptions, viruses you'll never catch, diseases you'll never get, and when needed, cures you never imagined.

This isn't a new aspiration. The *Huangdi Neijing* (黄帝内经), compiled over two millennia ago, is the foundational text of Chinese medicine. Like the *Tao Te Ching,* which we discussed in Chapter 5, "Nuclear War and Weapons," it represents core Taoist philosophy and it clearly understood the medical hierarchy. Known as *The Yellow Emperor's Inner Canon,* the treatise reads in modern translation: "Superior doctors prevent the disease. Mediocre doctors treat the disease before it is serious. Inferior doctors treat the disease after it has developed."

Examples abound of the ways current medicine, despite important strides, operates largely at the "inferior" level, treating diseases after they develop. Superficially, this is hardly a revolutionary insight. As a modern movement, what we now call preventative medicine traces to the 1964 U.S. Surgeon General's warning on the dangers of smoking. It caught on as a concept in the 1980s, when practitioners and insurance companies began to encourage and incentivize annual physicals and procedures like scheduled mammograms and colonoscopies. As it has since become standard in medical parlance, the core idea has certainly gained traction, along with the related concept of "holistic" medicine, which treats the person as a whole system of mind, body, and spirit.

As an example, the Dell Medical School founded in 2012 at my alma mater the University of Texas in Austin, is the nation's first modern medical school built from the ground up with a central mission of prevention and community health. I joined the Advisory Board of the Dell Medical School in October of 2025 as they race ahead to create the most advanced teaching hospital in the world, fully embodied with AI, sensors, and robotics since its beginning. With Neuralink and Paradromics nearby, I'm sure they will embrace brain-computer interfaces too, completing the Superfecta in healthcare.

While institutions like Dell Medical School represent a noble effort to stem the tide, the countervailing forces are formidable. Industrialized diets, fast food marketing, and vested lobbies that, combined with a fragmented healthcare system, cost Americans twice what citizens pay in comparably developed nations—while delivering shorter lifespans and higher rates of preventable disease. In a sick care model, the primary economic driver is the value of volume. Success is measured by how many hospital beds are filled, how many tests are administered, and how many prescriptions are written. It is a system that profit-maximizes on treating illness rather than ensuring health.

To reach the Age of Abundance for All, we must transition to a healthcare system based on the value of health. This isn't just a moral shift; it's a structural one. We need an "operating system" that rewards the prevention of disease with the same fervor that the current one rewards its treatment. The country of Bhutan, noted in the first of the *New Models for Humanity* in Part Three, "Transformation," is a good example. We'll explore there how Bhutan's citizens, despite their country being among the world's poorest by conventional measurement, still outlive Americans on average.

As the *Huangdi Neijing* reminds us, the goal isn't to be the best at fighting fires but to be the one who ensures the fire never starts.

Vast Disparities Define Health in America

Distressed communities, often Black or Hispanic, suffer a disproportionate share of this burden of illness. In my beloved state of Texas, a 2022 study by the Episcopal Health Foundation found vast health disparities co-related with income. In my hometown of Austin, it found a seventeen-year gap in life expectancy between the residents of privileged West Austin and the less privileged neighborhoods of East Austin. Diet, life stress, exposure to industrial toxins, and even environmental factors such as tree cover are the acknowledged culprits. This doesn't mean we should ignore drugs and other treatments, of course. The fast-lane science that yielded the mRNA COVID-19 vaccine in less than a year in 2020 saved millions of lives, but that speed was the exception and far from the rule. The pipeline for new treatments is slow. In the last thirty years, more than 200 proposed treatments for Alzheimer's have failed in clinical trials. While some promising treatments for depression—including psychedelics, which we'll discuss in more detail in Chapter 12, "What Is Our Reality?"—are in advanced trials, most treatments are still based on the mechanisms and assumptions from the 1950s. Amid the deadly rise in pathogens resistant to antibiotics, which kill more than 60,000 people in the United States alone, the most recent breakthrough is a drug called teixobactin, discovered in 2015. As of this writing in early 2026, teixobactin has still not even entered human clinical trials.

In the next chapter, "The Objectification of Women," we'll discuss in detail the political and social challenges women have faced—and overcome—throughout history despite many hurdles remaining. But I want to underscore here how this is particularly the case in our medical system, which is woefully inadequate at researching, understanding, and treating women's health issues. Piraye Beim, founder and CEO of Celmatix, a leading innovator in precision medicine for women's health is one of the world's leading pioneers who is confronting this in dramatic ways. Piraye, a good friend whose path has crossed with mine many times, was my guest for episode 31 of the *Love Conquers Fear* podcast. She explained not just the daunting problems, but the planet-changing innovations that she is bringing to women's healthcare—largely through AI. With the development of unique datasets specific to women's anatomy and biology, and by tapping neglected resources like Indigenous knowledge, traditional medicine, and alternative modalities, Piraye is collapsing all kinds of barriers.

"You would think just from a place of sheer curiosity, discovery, and a desire to go after white space and unlock these hidden mechanisms of our biology, pregnancy and the female body would be the most studied aspect of biology, not the least studied aspect of biology," Piraye told me. But she then explained how her company is reversing that: "On a very small budget with a very small skeleton team, at a very under-resourced little women's health company, we basically cracked open the

Holy Grail that pharma has not been able to do with all their money and all their time, thanks to AI."

The fragility of our healthcare system is also seen in its secondary effects. The decades-long catastrophe of opioid addiction and deaths, a crisis uniquely manufactured by doctors and pharmaceutical companies, has killed 500,000 people since 1999 and little effective treatment exists—yet. The epidemic number of suicides among military veterans—more than 40,000 since the events of 9/11 and quadruple the number of soldiers killed in combat in Iraq and Afghanistan—is a grim tally and one that grows by seventeen vets *each day* as an effective treatment for PTSD eludes us.

Those numbers moved me to host a fundraising dinner in September 2025 for former Texas Governor Rick Perry and Bryan Hubbard, CEO of the Americans for Ibogaine initiative, of Kentucky. Texas recently passed SB 2308, which allocates $50 million and will seek additional matching funds for Ibogaine clinical trials. This holds promise to be one of the most effective treatments ever for the scourges of opioid addiction and post-combat PTSD. If you haven't seen the moving documentary, *In Waves and War*, it covers this in detail. Also, Bryan Hubbard covered the sad state of opioid statistics as his big why on episode 53 of the *Love Conquers Fear* podcast. The funding comes amid a broader surge in private and philanthropic funding for psychedelic research—over $350 million was invested globally in the psychedelics sector in 2025 alone—with Texas's allocation the largest public investment in the field to date.

I digress a bit here, but it is important to mention as we'll be discussing this topic deeply in Part III, "Transformation." Ibogaine is the chemical derivative of the active component of the *Tabernanthe iboga* plant out of West and Central Africa, known for its healing powers from ancient tribal knowledge and research. It is another illustration of how much humanity has lost by consigning ancient technologies to the category of the "primitive"—and with them, the brilliant and gifted minds who developed those resources over millennia. The label, as it happens, says more about the labelers than the labeled.

The strides come while still, on average, it takes more than a decade and billions of dollars for a single drug to reach approval, and most fail along the way. The obstacles are many, but one key problem is economic: With R&D costs in the billions, companies are incentivized to focus first on drugs that serve the largest markets.

This paradox of elusive illness prevention, foreshadowed by ancient wisdom, is perhaps best illustrated in the skyrocketing popularity and use of GLP-1 and other weight loss drugs such as Ozempic, estimated by Morgan Stanley to have generated $15 billion in global sales in 2024, with projections heading toward $150 billion in 2035. To be sure, for those battling obesity and associated diseases like diabetes, these are life-saving drugs. And if they help us reduce the $173 billion in obesity's annual costs to the U.S. healthcare system, we will all rejoice.

Yet the counsel of *Hangdi Neijing* today would direct us to the root problem. In 1960, fewer than 13 percent of Americans were obese; since then, rates have tripled and today more than 40 percent of American are obese, according to the CDC. Imagine if we had prevented the obesity epidemic rather than spending billions to treat its symptoms. Those resources might have funded breakthroughs for children facing other devastating rare diseases—from muscular dystrophy to cystic fibrosis to childhood cancers—conditions that devastate thousands of families each year. Or imagine further that we could do both: really prevent such epidemics as obesity from occurring in the first place while accelerating treatments for *all* illnesses, even those diseases unlikely to yield an impressive return on investment.

And let's leap to another possible accelerant: the process of secondary approvals. Ozempic is a good example of how many drugs are discovered to serve other purposes beyond those first intended. Originally approved by the FDA in 2017 for the treatment of Type 2 diabetes, physicians soon noticed Ozempic's striking effects on weight loss. In 2021, the same active ingredient, semaglutide, was approved under the brand name Wegovy for chronic weight management. Now, further research suggests the drug may also improve metabolic and cognitive function— potentially benefiting patients with Alzheimer's disease—and clinical trials exploring these applications are underway. It's a topic I discussed in detail with author, biohacking expert, and good friend Ben Greenfield on episode 34 of the *Love Conquers Fear* podcast. Developing new medications and discovering the full range of their applications should take weeks or months, not years or decades, and making that a new reality is the promise of AI.

Rise of Obesity in America: 1960-2023

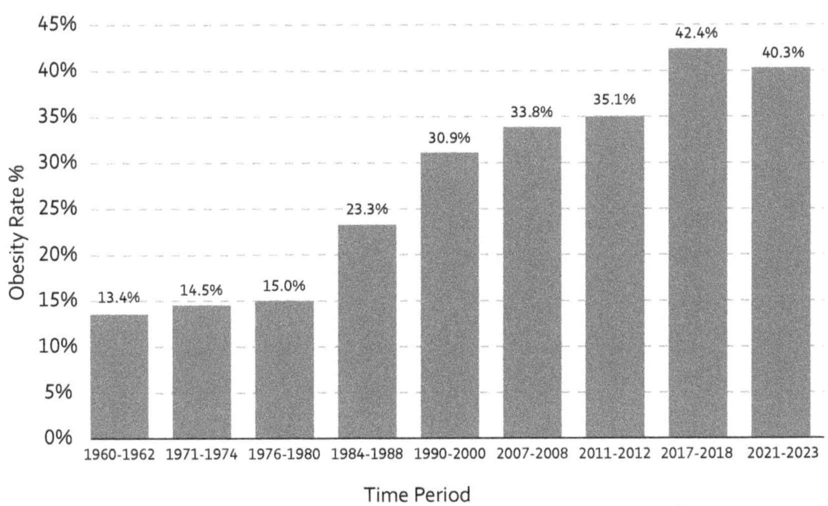

Image 7.1: Data compiled from U.S. Centers for Disease Control and Prevention.

Few have studied the cause and implications of our general tendency to treat healthcare akin to the "inferior doctors" called out in ancient Chinese medicine with the rigor of my good friend and fellow Austinite, Jason Karp. Jason is the founder and CEO of HumanCo, a holding company investing in a broad array of healthier products. Jason joined me on episode 3 of the *Love Conquers Fear* podcast to discuss the big problems he sees in the U.S. food system. We allow, he noted, more than 10,000 ubiquitous ingredients in our food that are banned in Canada and Europe. We spoke of industrial processes in agriculture and much more.

Diet is not the only culprit, of course. Depression is also skyrocketing, surging 60 percent just in the last decade, according to the CDC.

In 2023, the Surgeon General declared a national epidemic of loneliness and social isolation in his report, "Our Epidemic of Loneliness and Isolation." Many have compared that report to the seminal warning on cigarettes in 1964.

"I believe today we are in a true slow-motion apocalypse. We are in what I call a meta-crisis: a crisis of physical health, mental health, and planetary health," Jason told me, comparing our physical, mental, and social health outcomes to other countries. "[In] Europe, and that includes France, Italy, Spain, Portugal, they live five to seven years longer than we do. They have significantly lower rates of diabetes, heart disease, cancer and, by the way, they smoke way more than we do. They're also much happier than we are, and then, most importantly, on all the key metrics they are far better than America."

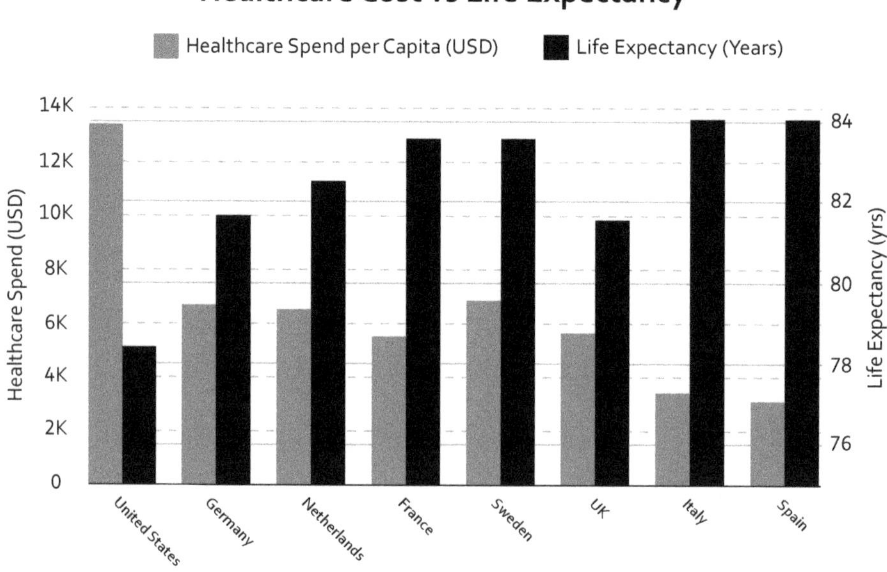

Image 7.2: Data compiled from OECD Health Statistics, Commonwealth Fund, and national health databases.

Jason, who overcame a diagnosis of incurable eye disease through his learnings and embrace of the medical practice and diets of Indigenous peoples around the world, is pushing the envelope of healthcare in many ways. A very successful hedge fund manager at the time, Jason began his new journey alongside his wife, Jessica, and brother-in-law, Jordan, with a health-focused restaurant in New York, Hu Kitchen. Since then, his new venture HumanCo has invested in companies ranging from Amara, which produces organic baby food without synthetic preservatives, to Grove Collaborative, a leader in sustainable home and personal-care products, to True Food Kitchen, a national restaurant chain centered on health-forward dining. HumanCo's portfolio also includes Against the Grain, a gluten- and grain-free food company, and Coconut Bliss, a pioneer in plant-based frozen desserts. We are proud investors in both True Food Kitchen and HumanCo.

Another friend in Austin who has placed himself at the center of this transformation is John Mackey, famously the co-founder and former CEO of Whole Foods Market. John, who wrote the Foreword to my first book, *The Entrepreneur's Essentials*, has influenced me in extraordinary ways, and he is certainly part of the inspiration for this book.

John, who sold his iconic network of stores to Amazon in 2017, is not slowing down his innovative pace. At seventy-two, he's well along in the next stage of his mission with the launch of his newest venture, Love.Life, a fusion of medical center and health spa that brings together western medicine with the ideas of *Huangdi Niejing*. "Love.Life called me forward, beckoning me back onto the field with a new opportunity to create, to risk, to play, to learn, to evolve, and to grow," he wrote in his brilliant 2024 autobiography, *The Whole Story*, which he and I discussed on stage at Austin's Jewish Community Center shortly after publication. He described the new venture as "Another chance to choose love over fear, to pursue innovation and contribution rather than settling for stasis and limitation." John and I discuss the state of the American diet, Love.Life, spirituality, and much more on episode 30 of the *Love Conquers Fear* podcast.

What Jason, John, and countless others are doing to detoxify our bodies and health system is urgent and vitally important. As Jason and I discussed on the *Love Conquers Fear* podcast, we are poisoning our minds with social media by hijacking our amygdala through the algorithmic seek-and-capture of our attention to maximize advertising revenue, as we discussed in Chapter 1, "Conquering Fear." Jason's point is that large food companies are doing the same to hijack our bodies into addiction by the constant innovation of chemical additives, which are mostly seen as poisonous and therefore illegal additives in most developed countries around the world. Yes, we still have a long way to go to meet these challenges at scale—toward the *"for All"* in the Age of Abundance. This is where the technologies of AI, synthetic biology, materials science, and robotics are imperative.

Preventative Health and Early Treatment at Global Scale

I'm not suggesting this will be an easy path with a straight line to perfect health. Without falling prey to calls to stop or slow AI development in general, and healthcare specifically, we need to achieve discipline and global sharing of best practices on a scale we've never managed before.

Mustafa Suleyman, the CEO of Microsoft AI and co-founder of Google's DeepMind—the lab that laid the foundation for our current progress in ways we'll discuss in a moment—framed it well in his brilliant 2023 book *The Coming Wave,* which I reviewed as, "A starting toolkit for humanity: Navigate the road to a future of AI-empowered Conscious Capitalism." The challenge is, as Suleyman put it, "Advanced AIs and synthetic biology will not only be available to groups finding new sources of energy or life-changing drugs; they will also be available to the next Ted Kaczynski."

Once more, among our tasks we must again lean toward ancient wisdom. "If one understands Heaven and Earth yet loses the Dao, even skill becomes perilous," counsels another passage in the *Huangdi Niejing*. In Chapter 11, "Risks," we'll explore more deeply the massive responsibility that must be shouldered by AI researchers and developers. This is, as I've contended in my science fiction novella *The Lattice* , God's final exam for humanity. But as we acknowledge and assume that responsibility, we must also not lose sight of the fact that for the first time in human history AI and robotics will allow us to truly move medicine from repair to prevention—not just for the fortunate few—but at global scale.

The road ahead is fraught with challenges. To Suleyman's point and that of many others, changing the course of the proverbial tanker—in this case the $5 trillion-a-year ship of American healthcare—is no mean feat. As futurist and AI pioneer Ray Kurzweil, among the biggest champions of AI in healthcare has pointed out, there's a great deal of understandable inertia in the medical establishment that is loath to change well established patterns and practices. "Nobody will want to be the person who approved a new and promising treatment on the chance that it turns out to be a disaster," he wrote in his 2024 book, *The Singularity Is Nearer*. I also reviewed that work when it was published, as "The AI-Driven Universe a Blink of the Eye Away."

Nonetheless, as Kurzweil points out, despite the cultural, legal, and regulatory hurdles, healthcare is where much of the early power of AI is being widely deployed and is accelerating—perhaps faster than in even some of the promising places we've explored, from energy production to education.

The World's First Drug Developed End-to-End with AI

In June of 2025, just as I began writing this book, Rentosertib, the first drug successfully designed end-to-end by AI, concluded phase II clinical trials for treatment of a rare lung disease. As of this writing, it is now set to begin phase III trials—the last stage before a new treatment can be approved for general use. This is a turning point that is hard to exaggerate. Much will follow when Rentosertib gets the final authorization from the FDA, which I believe it will.

As Kurzweil frames this juncture: "When medicine relied solely on painstaking laboratory experimentation and human doctors passing their expertise down to the next generation, innovation made plodding, linear progress. But AI can learn more from data than a human doctor ever could and can amass experience from billions of procedures instead of the thousands a human doctor can perform in a career."

This is how we are progressing. The pivotal moment, building on the work of thousands of scientists and decades of research, came with the work of Google's DeepMind just in 2020.

It's important to keep in mind that technological innovation moves in incremental ways, but is suddenly propelled by breakthrough moments that inspire scientists, motivate investors, and galvanize public opinion. Just as on November 30, 2022, ChatGPT riveted the world's attention on the remarkable speed and scope of generative AI's large language models, or LLMs, of OpenAI, Anthropic, Google Gemini, and others, AI in healthcare had its watershed moment as well. This turning point was the 2020 breakthrough of Google's DeepMind research lab—the protein structure prediction technology known as AlphaFold. That breakthrough earned DeepMind CEO Demis Hassabis and his lead developer John M. Jumper the 2024 Nobel Prize in Chemistry. "[O]n average, it takes, you know, ten years and billions of dollars to design just one drug," Hassabis said on CBS' *60 Minutes* program in mid-2025. "We can maybe reduce that down from years to maybe months or maybe even weeks. Which sounds incredible today but that's also what people used to think about protein structures. And it would revolutionize human health, and I think one day maybe we can cure all disease with the help of AI."

Let's hover over that quantum leap for a moment.

The Great Codex of Modern Biology——AlphaFold

Proteins, often called the "building blocks of life," are the worker molecules within our cells. Reflecting the importance of diet, proteins derive from what we eat, and depending on the three-dimensional structures they assume, proteins essentially carry out all the tasks of our metabolism. Before they undertake these distinct tasks, proteins "fold" into three-dimensional, mission-specific shapes. Structural proteins like collagen create the strength and elasticity of our skin. Enzyme proteins catalyze the chemical reactions in our bodies, converting the alcohol in a beer you might consume into toxic acetaldehyde—which causes your pounding hangover headache in the morning—and then into harmless acetate that your body can use for energy. Transport proteins such as hemoglobin move oxygen and nutrients to where they are needed on your morning jog and send waste to where it needs to go to be expelled.

While biochemists might nitpick my analogy, I liken proteins to the Japanese form of paper art known as origami. If you fold the flat, two-dimensional paper correctly, it can become a three-dimensional crane appearing to fly, or a flower in a stage of bloom. If you mess up, you have just a crumpled paper, you get what my youngest child, Yuzu, may call an "anatomical *orizombie*."

These misfolded proteins, "orizombies," are the silent architects of our most devastating diseases, from Alzheimer's to Parkinson's. For fifty years, the "protein folding problem" was the ultimate locked door in biology; we knew the paper was being misfolded, but we couldn't see the hands doing the work.

AlphaFold is the universal instruction manual we've been waiting for. It hasn't just predicted one structure; it has provided the guide for nearly every protein known to science. By allowing us to see exactly how these "orizombies" form, AI is giving us the tools to move from treating the wreckage of a disease to redesigning the origami of life itself—turbocharging disease research, with more promising breakthroughs such as new drugs now in clinical trials.

AlphaFold is the great atlas of modern biology—a living encyclopedia of molecular knowledge. DeepMind keeps the database freely accessible to researchers while its spin-off, Isomorphic Labs, turns the technology toward commercial drug discovery. This open-source revelation now powers both academic science and a commercial engine. The vast library of predicted protein structures remains freely available, even as extended versions model not only proteins, but also their interactions with DNA, RNA, and the small molecules that form the basis of medicine.

Through partnerships with pharmaceutical companies, and

integration into Google's own cloud and AI platforms, this new generation of AlphaFold systems is being harnessed to design drugs, diagnose disease, and guide molecular engineering. In effect, DeepMind is translating the Codex from pure research into applied medicine. That remarkable breakthrough has become an evolving grammar for how life builds itself *and* now how it might be *repaired.*

It is against this backdrop, and those of turning points like the mapping of the human genome completed in 2003 and the subsequent breakthrough in 2012 when science perfected the gene-editing technology CRISPR, that AI is propelling us forward through all the challenges I've described.

We may barely be noticing many developments, but if you've had a mammogram or colonoscopy or MRI scan recently, the odds are more than good that the initial results were generated by AI, with your doctor confirming the diagnosis afterward. There's so much more that's uncommon now but will soon be commonplace. This will include procedures such AI-assisted modes of surgery, including "digital twin" surgery where virtual replicas of your anatomy, physiology, and even whole organ systems will guide surgeons as they operate on the real you. What we'll see, within perhaps a decade, are the scarcely imaginable ways we'll be eliminating disease with nanobots that selectively repair damaged cells or artificial organs that outperform those you were born with.

In the first instance of AI in current use, primary care physicians are increasingly using AI chatbots like Babylon and Ada to triage symptoms, giving patients faster initial guidance. Wearables such as Apple Watch and Fitbit have normalized continuous heart, sleep, and activity monitoring, offering physicians streams of real-time patient data. In surgery, robotic-assisted systems like the da Vinci platform are already widely deployed, improving dexterity and outcomes in minimally invasive procedures. These technologies may feel invisible, but they are woven into the current fabric of modern healthcare.

In her remarkable account of her journey to found the Stanford Institute for Human-Centered artificial intelligence, *The Worlds I See—Curiosity, Exploration, and Discovery at the Dawn of AI,* computer scientist and entrepreneur Dr. Fei-Fei Li, describes not just the past and future of

AI, but how much is happening as you read this page. It's a really moving book that I reviewed as "Attention AI technologists, it's time to merge intentionality with philosophy, ethics, and law." One example from Li's work is the suite of technologies her team developed and dubbed "ambient intelligence" to monitor hygiene and the pathogens leading to the hospital-acquired infections that kill as many as 100,000 patients each year in the U.S. and perhaps more than one million worldwide. Such ambient intelligence equips rooms with sensors and smart speakers to monitor sleep, respiration, and heart rhythms without contact, alerting caregivers before crises develop. "AI won't replace doctors, but doctors who use AI will replace doctors who don't," Li said in one widely repeated remark.

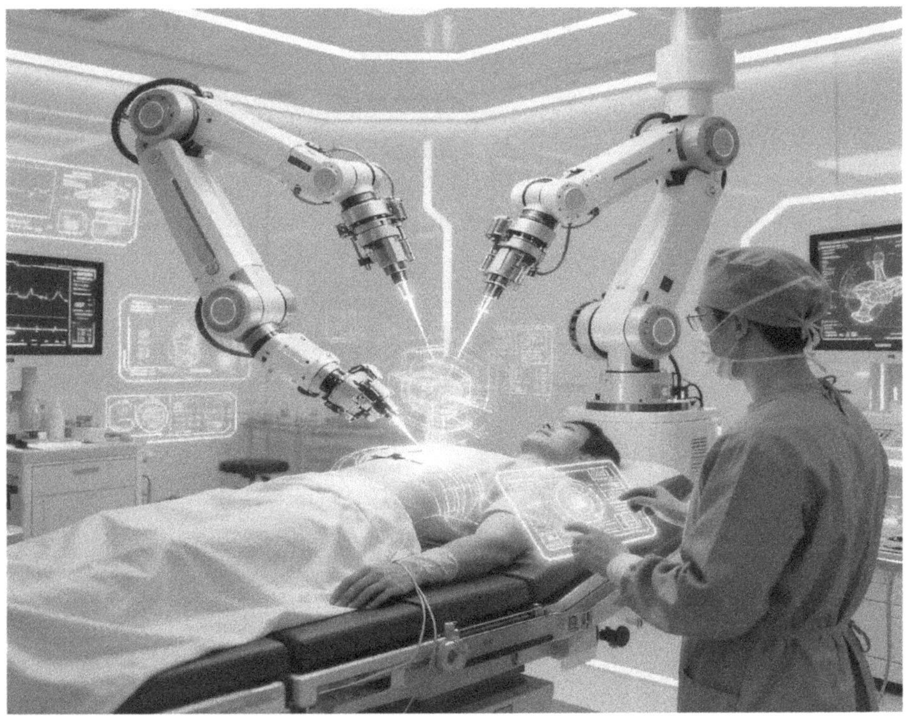

**Image 7.3: The future of surgery as imagined with AI.
LCF research using Perplexity AI.**

More distant but near, AI-assisted robotic surgery is advancing rapidly, with the digital twin technology I mentioned showing real promise. Early trials show that AI-driven robotic procedures can reduce operative time by 25 percent and complications by 30 percent. Personalized mental health support apps, powered by AI and linked to wearables, are moving toward mainstream adoption. Within a few years, precision imaging, robotic-assisted therapy, and AI-enabled telehealth will be as standard as today's MRI.

Looking further ahead, nanomedicine promises fleets of microscopic robots that patrol the bloodstream, repairing tissues at the cellular level. Synthetic biology and immunomics, the interdisciplinary study of diseases and the immune system on a genome-wide scale, point to a future where engineered cells preemptively block diseases before they manifest.

Advances in genomics powered by AI are expected to deliver true precision medicine, with therapies customized for each individual's molecular profile. Fully autonomous AI health assistants, essentially carrying out the ancient vision of *Hangdi Neijing,* will deliver anticipatory care—predicting illness before symptoms appear. Artificial organs and bioengineered tissues may outperform natural ones, as Kurzweil has predicted, offering new cures for organ failure. And AI-driven drug discovery pipelines may collapse the years-long process of developing therapies into months, unlocking cures at a scale we can barely imagine today. Yes, the challenges we face in healthcare are massive, as we know from interminable debates, shortfalls in coverage, anger and fear over vaccine policies and rules, and even a shutdown of the U.S. government over healthcare as I was writing this chapter. But we have the tools, the technology, and certainly the insight to actually turn this massive health complex around and steer it in a new direction. That direction must be toward the Age of Abundance for All.

Any intelligent fool can make things bigger, more complex, and more violent. It takes a touch of genius—and a lot of courage—to move in the opposite direction.

—E.F. Schumacher, economist and author of *Small Is Beautiful: Economics as if People Mattered*

If the first woman God ever made was strong enough to turn the world upside down all alone, these women together ought to be able to turn it back, and get it right side up again! And now they is asking to do it, the men better let them.

**—Abolitionist Sojourner Truth,
in her speech of 1851: "Ain't I a Woman"**

Of all the cultural forces currently reshaping our world, few are as misunderstood—or as vital to the Age of Abundance for All—as the global movement for women's liberation. We must look past the narrow political skirmishes of the last century to see this for what it truly is: an engine of awakening.

Women's liberation is not merely a "women's issue" that too often gets shortsightedly labeled as feminism; it is a fundamental shift in the human operating system. It is the counter-force to the zero-sum power dynamics that have fueled the meta-crisis of war, environmental degradation, and social alienation. When we speak of the liberation of women, we are speaking of the liberation of the human spirit from the constricting confines of patriarchy that have adversely affected not only women but also our sons. This has led to disparities in educational performance and a pervasive crisis of purpose among men. From the ancient "superior doctors" who strove to prevent the disease to the modern leaders striving to keep peace, the feminine principle offers the essential perspective needed to navigate the turbulent river of our future.

This is not to diminish terms like feminism or liberation nor the vital struggles that women's liberation has undertaken. Rather, the ethos animating women's struggle for equality—from Seneca Falls in 1848, to the ratification of the Nineteenth Amendment in 1920, to the #MeToo sparks of 2006 that fired into a global movement a decade later—is the engine of humanity's awakening. Soon to be spurred on by AI and the technologies we've been discussing, this engine is driving us toward the Age of Abundance for All, when love will become our common bond. "Feminism is for everybody. Passionate politics is for everybody. The soul of feminist struggle is the commitment to end domination in all its forms," wrote the author bell hooks (who spelled her name all lowercase) in her 2000 book, *Feminism is for Everybody.* In another book that deeply informs this one, *All About Love: New Visions*, also published in 2000, hooks explored the intersections of race, class, and gender in shaping systems of oppression, writing:

> *To know love we have to tell the truth to ourselves and others. Creating a false self to mask fears and insecurities has become so common that many of us forget who we are and what we feel underneath the pretense. Breaking through this detail is always the first step in uncovering our longing to be honest and clear.*

We'll return in this chapter to the history of women leading humanity toward freedom, and to hooks' conceptualization of love as not just a sentiment, but a muscular source of action. In her work hooks emphasized love's utility not as a noun, but as a verb. We'll also explore the inspiration to be found in the ancient "Two-Spirit" concepts of third and even fourth gender identities that existed across many Indigenous North American cultures and the role of the transgender Hijra communities of South Asia that today illuminate the truth that human wholeness requires fluidity and reciprocity.

The "De-Objectification" of History

As we saw in the prior chapter with Piraye Beim and her work at Celmatix, the "white space" of women's health is the ultimate proof of our meta-crisis. It is a staggering reality that the female body remains one of the least studied aspects of human biology, which is a direct result of centuries spent treating women as *objects* of medical study rather than as the primary *subjects* of their own lives. For example, the disease endometriosis is not only debilitatingly painful for young women who have it. They miss volleyball practice, drop out of the lacrosse team, and struggle through their formative years, but the impact doesn't end there. Later in life, endometriosis sets women up for mental health issues and infertility. When they are able to get pregnant, they face greater risk of dangerous outcomes like preeclampsia. And lifelong, women who have endometriosis carry a higher risk of certain cancers and heart disease, the number one killer of women.

This medical blind spot is the first layer of objectification we must strip away. When we ignore the unique anatomy and biological data of half the population, we aren't just being unfair; we are being unscientific. "It's such an impactful disease," Piraye said of the malady endometriosis as a guest on episode 31 of the *Love Conquers Fear* podcast, "yet we spend less researching it than we do on male pattern balding." By leveraging AI to finally map this *terra incognita*, we do more than just cure diseases; we begin the process of "de-objectification," and begin restoring agency, authority, and medical sovereignty to women as we move toward an Age of Abundance that is truly for All.

To consider how we got here, we must look boldly at the

immediate pathology before us: a culture where a "locker-room" mindset of male dominance over women was not only excused but elevated to the highest office in the country. This shift has done more than poison a generation of young men; it has imprisoned our entire culture in a paradigm that treats human beings as objects to be "grabbed" rather than fellows to be respected.

I realize that for some, this critique may feel like the preamble to a "woke" lecture, but before you slam this book shut, understand that my perspective isn't rooted in a partisan identity. As an independent active in centrist organizations like No Labels and the American Israel Public Affairs Committee, also known as AIPAC, my commitment is to bipartisanship and national unity. Both No Labels and AIPAC are rooted deeply in bipartisanship and focused on bringing our country together, not dividing us like the far left and far right try to do—a divisive quest the media and social media algorithmically assist. I've freed my mind (thank you, Morpheus) to call balls and strikes based on a simple standard: Does a behavior lead us toward the Age of Abundance for All, rooted deeply in love, or does it keep us trapped in a scarcity mindset rooted in fear and domination?

This is in some ways a subset of the anxiety we explored in Chapter 1, "Conquering Fear," but it is a trend with its own dynamics. The crisis of toxic masculinity that now grips so much of the world is not an isolated dysfunction; it is the visible symptom of patriarchy's collapse. Across societies, young men drift between rage and despair, seduced by digital echo chambers that promise belonging through domination. We witness the loneliness of boys who vanish into the virtual world of the "manosphere" and emerge as bullies, even vectors of the scores of school shootings and political violence endemic in America. As I write this, I was just at the Austin City Limits Music Festival in Austin, where Sabrina Carpenter performed. Her most popular new song is "Manchild" on her wittily satirical album *Man's Best Friend*.

We see the phenomenon of *hikikomori* of Japan—predominantly young men who withdraw from social life and remain isolated in their homes, sometimes for years. Once thought to be a uniquely Japanese pathology, with numbers estimated in excess of one million, this phenomenon has now spread to other Asian countries such as South Korea, China, Thailand, Singapore, India, and Bangladesh. In the Middle East, we see radicalism emerging not just in the struggling societies where youth unemployment often exceeds 50 percent, but in middle class or even wealthy homes, such as those which produced the 9/11 hijackers—most of whom were educated men in their twenties from middle-class or affluent backgrounds—several of whom had university degrees.

This viral pathology is hardly confined to just the young. It is effectively guided and mentored by a brutish overclass that we see in the demagogic swagger of strongman leaders in America, Turkey, Russia, India, Hungary, Venezuela, El Salvador, Argentina, and among those on deck awaiting their turn in France, the Netherlands, Germany, and many other places, all nurturing hatred as a form of political capital. This overclass often invokes mythic figures from the past: the combative Andrew Jackson of nineteenth century America, the heroic Rama of India's epic kingdom before the first millennium, and the new veneration of fifteenth century Ottoman conquerors by today's leaders in Turkey.

This should not surprise us. Today's atomization of people into lonely subjects, willing to accept lies and be swept into ideologically sanctioned violence, was foreshadowed from history long ago by author and Holocaust scholar Hannah Arendt, whom you met at the outset of this book. These are all variations of the same festering wound—a hunger for connection, agency, and purpose—that has been twisted into aggression and all too often manifests in violence.

Scores of books, analyses, and commentaries, certainly enough to fill a small library, have explored the origins and causes of this malaise and accelerating anxiety among men, particularly young men. Three decades ago, poet Robert Bly, who inspired the "mythopoetic men's movement" saw the culprit in changing gender roles. While he explicitly claimed to welcome the rights and freedoms women achieved in the 1960s and 1970s, he also saw this achievement, quite controversially, as part of an usurpation of inherited male identity. His 1990 book, *Iron John: A Book About Men*, uses the nineteenth century Brothers Grimm folk tale "Iron John" as a psychological allegory of repressed masculine energy that must be recovered, freed, and integrated to transform what he called "perpetual adolescents" into mature men. "Men and women alike once called on men to pierce the dangerous places, carry handfuls of courage to the waterfalls, and dust the tails of the wild boars. All knew that if men did that well, the women and children could sleep safely," Bly wrote in this lyrical book. "Now the boars have turned into pigs in the stockyard, and the rushing rivers to the waterfall into the Museum of Modern Art courtyard. The activity men were once loved for is not required."

Bly's is an interesting argument, and his insights on the importance of initiation rituals are convincing. Whether they be the traditional B'nai Mitzvahs of Judaism, Eagle Scout badges on the passage to manhood, or fraternity rites on college campuses, these topics resonate with many others, including the entrepreneur and prolific writer Scott Galloway, upon whose work I'll again touch on in a moment.

My concern with Bly is not that he is wrong. His book, after all, was a cultural phenomenon. With more than a year on the *New York Times* bestseller list, it spawned countless men's retreats, drum circles, and workshops. Rather than wrong, I find it incomplete as it skirts vast swaths of the challenge. It is incomplete first in that it is specific to the American context. It is further incomplete in an evolutionary sense as it frames the emergence of women's power and men's reaction as a binary tension when in fact it's better thought of as a circular ring of tension, a yet to be completed but ultimate wholeness.

I realize I'm breaking new ground for many readers here, so bear with me. For example, in a very different cultural context, some commentators have speculated with a mythic interpretation akin to Bly's, that Japan's contemporary pathology of male withdrawal can be read against two distant backdrops: first, the 1950s to early 1960s student mobilization against the U.S.–Japan Security Treaty and remilitarization, and later, the sense of anxiety and loss that followed the collapse of Japan's economic boom in the 1990s and the rise of precarious work.

From Japan, it's a short geographical and intellectual leap to examine another manifestation of young male alienation: the radicalization of some Muslim youth. Domestic political repression leaves little space for peaceful activism, argues University of Maryland psychology

Professor Arie Kruglanski in his 2025 book with Sophia Mosalensko, *The Psychology of the Extreme.* As a result, violence is sometimes reframed as the only available channel of resistance.

This occurs among both European-born young men, caught between two clashing cultures, and among their contemporaries in parts of the Middle East where male youth unemployment often exceeds 50 percent. We see how one group in Japan withdraws inward as another in Europe and the Middle East turns outward toward violence, while the underlying emotional terrain is strikingly similar: feelings of insignificance, humiliation, and invisibility. Radical ideologies, whether they lead to hermetic isolation or violent extremism, can appeal as a means to restore a sense of worth and purpose.

I share this broad sweep of history and geography to illustrate the breadth of the challenge, which is no less vivid and urgent in America. No one has explored this dilemma at home more effectively than Galloway, a writer, podcaster, investor, and business professor who riveted those attending the 2024 TED Conference, including me, when he rhetorically asked: "Do we love our children?" In this compelling talk , Galloway described just how much financially worse off are upcoming generations compared with earlier generations and how so many policies are effectively intergenerational theft. But he does not stop at the harsh economic realities. In his blog to which I subscribe and in countless interviews on the BBC and elsewhere, Galloway has described the problems of young men in even starker terms. He calls the situation a systemic "war on the young," drawing on data to show boys falling behind girls in education, men behind women in employment, and spreading loneliness. Galloway shows how boys are falling behind in nearly every metric—not just in education but in propensity to drug addiction, imprisonment, and death by suicide. This decline, he contends, threatens not only young men themselves but the broader social fabric. I agree.

For Galloway this malaise led to a breakdown of institutions that once served as stabilizing forces: schools that fail to accommodate boys' slower cognitive development, the decline of mentorship and vocational programs, and an economy that increasingly rewards capital rather than labor.

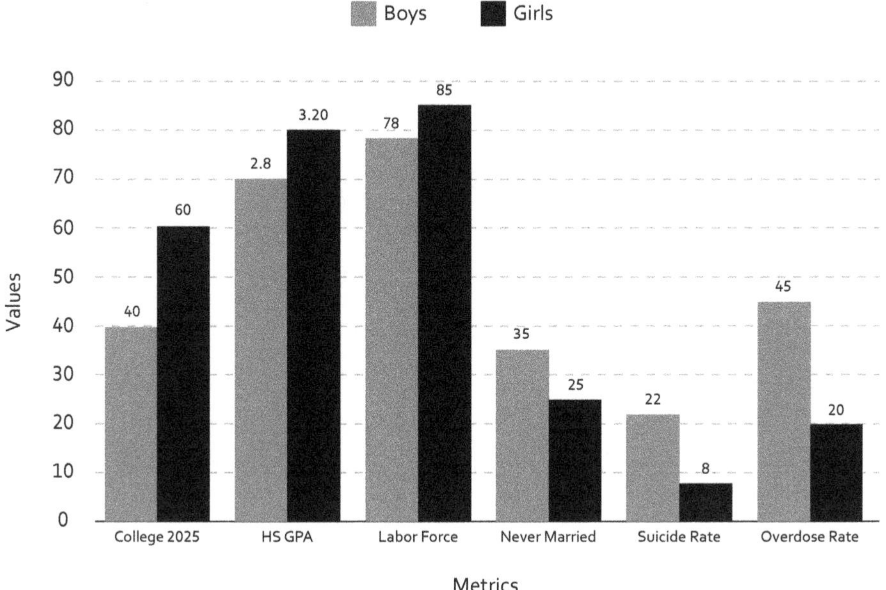

Image 8.1: Girls lead in percentage comparisons in education, early work, and marriage. Suicide and overdose comparisons are of incidents per 100,000 of population. LCF research using Perplexity AI.

Like Bly, Galloway sees the absence of male initiation rites that once transformed boys into responsible adults as a core cultural wound. Galloway's solution is pragmatic rather than Bly's nostalgic quest to recapture the past. Galloway wants public reinvestment in education, apprenticeships, and "third places" such as community groups and national service programs that foster purpose and belonging. He even proposes "redshirting" boys by delaying school entry by a year to account for developmental differences and improve long-term outcomes.

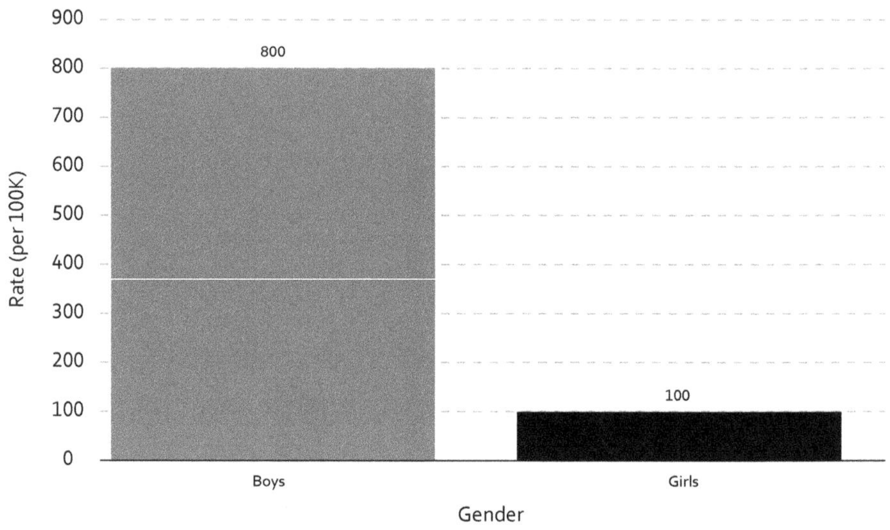

Image 8.2: Incarceration rates are 1 per 100,000 ages 20–29. LCF research using Perplexity AI.

The case Galloway argues is not just urgent in America, much of his diagnosis applies as well to the broad brush of deep and global problems we've discussed that stem from a masculinized, patriarchal reality. I share, support, and applaud his call to action—with a caveat. My caveat is that these gaps will only worsen if we do not come to better terms with the reality that the differences between boys and girls and men and women (and there are real differences) are not born of competition but of incompletion. This will only grow more vivid in the emerging new economy of AI.

The fear that AI will eliminate jobs is a topic of daily conversation and headlines, but what most of those projections and lamentations miss is the fact that AI and robots eliminate tasks. They do not eliminate relationships. Why is this important? It is crucial because human-centered, relationship skills are what will be in demand. Empathy, trust, communication, collaboration, and tolerance of ambiguity are skills that our paternalistic culture has devalued, meaning they are roles dominated by women. Think of teaching, healthcare, speech pathology, or counseling. In my world of technology, I think of customer success, product management, or human resources. Amid the current sense of urgency to ensure that AI is safe and responsible, the rising need is for skills in risk assessment, compliance, safety review, and stakeholder engagement—all skills demanding cross-functional collaboration and all competencies where women often outscore men.

The winners in this new moment won't be those who "know everything." Rather, the winners will be those who learn fast, ask good questions, or frame good prompts. Skills emphasized in many girls' education, study habits, meta-cognition, and feedback-seeking are skills that map directly to the leveraging of AI.

I don't have all the answers of course. And much of this goes back to Chapter 6, "Education" and our discussion of the problem of "linearalism," the pattern I discussed of two-dimensional solutions to three-dimensional problems. Girls do mature faster than boys in many ways, as all parents know. Yet we insist, outside of some experimental models, that all students shuffle along the same linear K-12 track with the same linear testing and grading systems.

Still, these are really tactical solutions short of the strategic shift in the paradigm of our worldview. At root to make this imperative shift occur, we must rethink the history of what we too narrowly label as a binary, which misconstrues the history of what we label as feminism or the women's movement.

To that point, this chapter's title, "The Objectification of Women," carries a double meaning. We typically understand it as reducing women to their bodies and sexual function—treating them as objects rather than as human beings. We've certainly mistreated women in health care research, as my friend Piraye points out. We've excluded women from clinical drug trials, denying us knowledge of how molecules interact with feminine biology. We've ignored many women-specific diseases such as endometriosis mentioned earlier. We have even built safety features into cars exclusively from the data gleaned from "male anatomy" crash dummies in testing. And this is despite the fact that women are now a majority of doctors in the United States, 65 percent of the students in medical school and some 80 percent of healthcare workers overall.

But there's another form of objectification we've barely acknowledged: reducing women's civilizational impact to a checklist of legal victories. Yes, the right to vote, own property, and receive equal pay matter immensely. But framing women's influence solely through these milestones obscures a deeper transformation: Women's participation has not only made societies more peaceful but holds the key to humanity's next leap forward.

"Could it be that radical change has gone unnoticed and unappreciated?" ask authors Joslyn N. Barnhart and Robert F. Trager of the pathbreaking 2023 book, *The Suffragist Peace—How Women Shape the Politics of War.* "Perhaps the changing political status of women—a process which is ongoing—has created the modern world more than we realize." International relations professors Barnhart and Trager peel back layers of women's social and political history that stand as empirical validation that feminism is, as hooks so elegantly phrased "for everybody," or as I would frame it, *for All.*

A compelling example of the universal importance of women's suffrage can be seen in that virtually all histories of the movement date its origins to the 1848 convention in Seneca Falls, New York, organized by Elizabeth Cady Stanton and Lucretia Mott, that passed the resolution leading to the Nineteenth Amendment seventy-two years later. That is a key moment in history, but lost to the popular memory of these events is the fact the resolution only passed because of the demands and compelling oratory of Frederick Douglass. Douglass was the former slave and towering intellectual better known for his contributions to abolition and another speech that I've reread many times, written nine years before the Civil War, "What to the Slave Is the Fourth of July?" This speech is required reading in the Aspen Institute's Henry Crown Fellowship.

"The most defining event of the conference almost did not happen," write Barnhart and Trager, because some women and many of the men in attendance felt that calling outright for the vote would be too radical. They continue, quoting Douglass, arguing for women that denying their access to politics is a form of degradation. The authors wrote that as the resolution for suffrage floundered, Douglass stood and eloquently petitioned the crowd, urging that the world will be a better place if women are brought into politics. Douglass told the gathering that denying women the vote equaled, "the degradation of women and the perpetuation of a great injustice . . . [the] maiming and repudiation of one-half of the moral and intellectual power of the government of the world." Without Douglass, the authors conclude, the resolution would likely not have passed.

Another insight lost when we stop at only the highest milestones is the role of women's political power before 1920, the year the Nineteenth Amendment was ratified, with the final statehouse vote coming in from Tennessee. That date is memorialized in the name of the national news organization, The 19*th*, based in Austin, that my wife Debra and I have supported since its founding in 2020.

That date of ratification is important, but how many of us are aware today that eleven states, all in the West, had already granted women the right to vote by 1916? To the point of women's unacknowledged power, it was the increment of that vote, Barnhart and Trager analyze, that narrowly swung the election to Woodrow Wilson who opposed the war with Mexico that Wilson's opponents Theodore Roosevelt and Charles Hughes were urging on the nation. While war in Europe did ultimately ensue under President Wilson, women's activism nonetheless contributed to the creation of the League of Nations, the forerunner of today's United Nations, and other Wilson initiatives, the authors conclude.

Another example goes to the suggestion made above that one contributor to the *hikikomori* malaise in Japan may be the country's deep ambiguity toward the pacifism imposed on the country by the U.S. after World War II. Repeated attempts have been made in recent years to repeal Japan's pacifist Article 9 of its constitution that would allow the country to reestablish a full military and rearm. The single reason successive governments have not taken this step is that it would require a referendum, which poll after poll indicates women would oppose, stopping the idea in its tracks.

The Hidden Sovereignty

When we strip away the lens of objectification, we find that feminine authority has never been absent; it has simply been ignored. From the clan mothers of the Native American Haudenosaunee, who held the power to appoint and remove chiefs, to the "superior doctors" of ancient medicine, women have long served as the architects of social stability. This is not "alternative" history; it is the reclaimed foundation of how we will govern ourselves in the Age of Abundance for All.

Too many other examples of women's influence abound in *The Suffragist Peace* for me to summarize, but they include the coalition of Muslim and Christian women who helped end Liberia's devastating fourteen-year civil war in 2003, for which leader Leymah Gbowee won the Nobel Peace Prize in 2011, and the grassroots "Four Mothers" movement in Israel, whose campaign contributed to ending the eighteen-year Israeli military presence in southern Lebanon.

The Suffragist Peace notes the question is often asked, "What if women ruled the world?" The book answers that query with what I think is a much better question: "What if political norms derived from women's voices, what sort of world would that be?" It's a great and enduring question, to which I would only add: What if political norms derived from the contributions of all genders—men, women, and those who choose to be both or neither?

At a time when federal executives, state lawmakers, and cultural warriors are falling all over one another to legally codify law and policy into "two sexes: male and female," science, biology, and medicine are far more nuanced. Embryos, for example, develop along a shared, undifferentiated path for six to seven weeks before sex characteristics form through active genetic and hormonal signaling. The clinical mismatch between identity and assigned sex, known as "gender dysphoria," is a real and well-studied clinical condition. And a small but very real number of people are born with differences of sex development, as "intersex," with conditions that can include tissue typical of both ovaries and testes.

Reality doesn't fit neatly into two boxes, a fact many societies have long recognized. From numerous Indigenous cultures in North

America to Hijra communities in South Asia, there are culturally specific ways of honoring gender diversity, embedding it in spirituality, kinship, labor, and ritual life rather than treating it solely as an individual identity claim. Our debates and policies can draw from these lessons to imagine political norms shaped by the full range of human experience.

Among Native peoples, including the Zuni, Diné (Navajo), and Lakota, gender-diverse people frequently took on roles in healing, mediation, ceremonial leadership, the preparation of ritual objects, and acted as go-betweens in diplomacy or family conflict. Beyond these Indigenous examples, the Hijra—a distinct socio-religious group of trans and intersex people in India, Pakistan, and Bangladesh—has been recognized for centuries and remains active today with diverse ritual authority. The Hijras are key at many life-cycle moments, such as childbirth and weddings, and offer blessings and at times feared curses—a combination that confers them with both social authority and power.

The Soul of the River

To conclude, "de-objectification" of women is among our many tasks as we approach the bend in the river of humanity, and gender is certainly where we will encounter both turbulent currents and white water. But to navigate these shoals we must not only pass beyond the objectification of women, we must also surpass the objectification of women's political history as limited to votes and ballots, and we must certainly overcome our gender phobia.

While he might not have intended it this way, a line from Robert Bly's *Iron John* speaks on this, suggesting this about our bend and metaphor of transformation from a Celtic story: "[It]is more appropriate to say the water is soul water, and as such, both masculine and feminine." For me, the river is both and it is neither, its spirit and souls are merely one.

I am not what happened to me,
I am what I choose to become.

—Carl Jung, psychiatrist and founder of
the concept of collective unconscious

CHAPTER 9
CAPITALISM

Nothing is enough for the man
to whom enough is too little.

—Epicurus, founder of
Epicureanism (circa 306 BCE)

In the history of human progress, there have been pivotal moments when the foundations shifted so fundamentally that we could hear the gears of civilization clicking into place. One of those moments arrived in late October 2025.

For years, the race for Artificial General Intelligence (AGI) felt like a private arms race—a high-stakes Silicon Valley game of "Capture the Flag" driven by the logic of scarcity and proprietary control. But on October 28, 2025, OpenAI took a bold step with its foundational pivot, signaling an acceleration toward the Age of Abundance for All.

In a singular, digital hat trick, OpenAI didn't just chase a $500 billion valuation; they reengineered their corporate DNA. By converting into a Public Benefit Corporation (PBC), they established a legal commitment to prioritize serving humanity above all else. Most significantly, they endowed the non-profit OpenAI Foundation with a staggering $130 billion—a substantial fund dedicated to addressing the "meta-crisis" of healthcare. And OpenAI followed this up shortly after with the launch of ChatGPT Health on January 7, 2026, followed by Microsoft launching Copilot Health just two months later on March 12.

This was more than just a restructuring; it was a declaration that the most powerful technology in human history would not be a private hoard, but a public utility. OpenAI effectively proved that the "superior doctor" model of prevention we explored in Chapter 7, "Healthcare" could be fueled by a "superior economy." It marked the moment when the soul of a new kind of capitalism became impossible for the world to ignore.

Take a bow, team OpenAI.

The company's creation, as we know, is a global game changer. But when someday the history of AI is written somewhere in the cosmos, October 28, 2025 may prove to have been as paradigm-shifting a moment as the launch of ChatGPT itself on November 30, 2022. Let's explore why this action represents a fundamental reorientation about ideas and outlooks on business and commerce. First, a bit of context on this painstakingly achieved milestone by decade-old OpenAI, with its huge meaning for the public benefit corporation movement—the foundation of the AI economy to come.

The Recentering of Capitalism

To understand why OpenAI's move is so revolutionary, we have to look at the legal DNA of modern business. For decades, the dominant operating system of capitalism has been shareholder primacy—a rigid code that mandates directors prioritize short-term stock prices, or shareholder value, above all else. In the Age of Abundance for All, this OS is obsolete. We are moving toward a new OS that will prioritize stockholder vitality, where the health of the community, the environment, and employees are seen as the fuel for long-term profit, not a distraction from it.

This movement manifests in two distinct but complementary layers. The first is the B Corp Certification. Think of this like the UL logo on your toaster; it's a voluntary, audited gold standard that tells the world a company meets high social and environmental performance standards.

The second, more foundational layer is the Public Benefit Corporation (PBC). This isn't just a rigorous certification; it's a legal steel cage. By incorporating as a PBC, a company like OpenAI or data. world bakes its mission *directly* into its corporate charter. It gives directors the "legal permission" to make decisions that serve the public good even if they don't maximize immediate profit. In an era when AI can create endless value, the PBC structure ensures that value is governed by human values, rather than solely by market mechanics.

Without getting carried away on the history, it's worth noting that the idea of what we now call stakeholder values, with antecedents back to the Great Depression, was popularized by author and entrepreneur Paul Hawken with his 1993 book *The Ecology of Commerce*, and further with Amory and Hunter Lovins' book *Natural Capitalism* in 1999, which was co-authored by Hawken.

This was followed by the Conscious Capitalism movement led by my good friend John Mackey, the co-founder and former CEO of Whole Foods. John spearheaded the creation of Conscious Capitalism in 2010, a non-profit organization to help companies adopt a higher purpose beyond mere profit. I served on its Board of Directors, and my last company, data.world, was both a B Corp and a PBC from the outset of its public launch on July 11, 2016. Two of John's books—*Conscious Capitalism* in 2013 and *The Whole Story* in 2024—furthered the ideas of principle and purpose, as I sought to do with my own 2022 book, *The Entrepreneur's Essentials.*

I'm deeply invested in these models because they represent the kind of institutional evolution humanity demands. OpenAI is certainly not the first PBC, and other AI companies—notably Anthropic, which as of this writing is gaining on OpenAI's valuation and prominence—are also PBCs. But for a company of the size, scope, and reach of OpenAI to embrace the PBC model is like a giant redwood sprouting amid a grove of vibrant saplings and sturdy oaks, reshaping the forest of capitalism and signaling the most profound transformation of commerce since the birth of the joint-stock company.

The "Re-Recentering" of Capitalism

An elegant way to think about the transformative power of AI is as the "third decentering of human consciousness," in the term of Jonathan S. Blake and Nils Gilman, in their book *Children of a Modest Star*, mentioned earlier. I wrote a review of the book on Thanksgiving in 2024 ✈, "How the Ethos That Led to the First Thanksgiving Should Guide Us in the New Age of Discovery." Their use of the term evokes the first decentering, when Copernicus moved Earth from the center of the universe, followed by Charles Darwin, who displaced humans from the pinnacle of creation. This third recentering forces us to recognize that we're not separate from planetary systems but embedded within them, a core belief of mine as we discussed in Chapter 3, "Geopolitics." Against this backdrop, even capitalism will require fundamental rethinking. What we are doing is recentering capitalism, along with many other things in the Age of Abundance for All.

Not that I'm a critic of capitalism. Quite the contrary. Capitalism has been extraordinarily effective at leveraging and managing scarcity—the state of humanity that we are leaving behind. Does capitalism have flaws? Of course it does, and critics include not just those familiar on the progressive left, but stalwart capitalists such as Ray Dalio, the billionaire founder of Bridgewater (the world's largest hedge fund), Mark Cuban, the billionaire investor many know from the TV show "Shark Tank," and many others. Dalio advocates reforms of the tax system, education, and business practice to confront the challenges of growing income inequality and the stagnation of working class wages over the last forty years. Cuban makes similar arguments, and has taken the initiative to launch such projects as his Cost Plus Drug Company. By cutting out middlemen and simplifying prices, Cuban's company can sell many pharmaceuticals, without health insurance, that are actually cheaper than even the co-pays at typical pharmacies. There is so much that we can and will do as the contours of a new economic equilibrium emerge.

But attacking the premise of capitalism is hardly the answer.

As Harvard's Steven Pinker put it in his profound 2018 book, *Enlightenment Now*, this engine of trade and commerce is responsible for most of the advances in health, life, and living standards for more than 200 years: "Capitalism, with its incentive for innovation and diffusion, has lifted billions of people out of poverty and improved the well-being of almost everyone on the planet," he wrote. Stop and consider for a moment that we've only had home electricity since its beginning in 1882 when capitalist Thomas Edison created the first power station. Flush toilets were not common until commercialized by entrepreneurs two decades later. Antibiotics were experimental until 1943, when U.S. pharmaceutical firms—notably capitalists Pfizer, Merck, and Squibb—developed the technology for mass production. This puts all of us in the top 0.1 percent of all Homo sapiens—meaning that with our evolutionary history of around 300,000 years, 99.9 percent of our ancestors went without these abundancy-triggering conveniences.

But now what? How will businesses profit from abundance? How do we tax labor when it's all done by robots? As we already recognize corporate "personhood" —the result of such Supreme Court decisions as Citizens United or Hobby Lobby —how shall we acknowledge this synthetic entity of AI when it begins to speak for itself? Who or what regulates a technology that can outsmart any regulator? What's the new scoreboard for life's meaning itself?

As Peter Diamandis, the visionary we met in Chapter 2, "Abundance," the originator of the concept more than two decades ago, asked in a 2025 Substack essay: "What happens when superintelligence, humanoid robotics, and nanotechnology drive production costs toward zero, eroding profit motives?" Diamandis envisions this as a "post-capitalist" society and concludes we've got a lot to figure out.

With abundance, Diamandis wrote, scarcity shifts to things that can't be replicated: experiences, creativity, and meaning. These are what will be truly limited. I heartily agree with that, though I'm not sure the phrase "post-capitalist" is the best framing. Yes, post-capitalist is a term increasingly used by those pondering the future, including influential economic theorist Jeremy Rifkin, the author of two dozen books, including the 2014 prediction of the eclipse of capitalism, *The Zero Marginal Cost Society.* But, we're not going to replace capitalism. As we've done in the past, we'll rethink and evolve it into new forms, as we've done countless times.

Rather than post-capitalist, I argued in a 2024 post in the wake of Sam Altman's brief ouster, "The 11 Lessons Three Months on from OpenAI's Five-Day Shakespearean Drama," the better term, is a "post-digital economy." I'm not suggesting that we abandon digital tools and technologies. Not at all. I do believe, however, that the familiar divide between analog and digital is obsolete. Soon, AI will be ushering in a world where digital technologies are—and will be—so embedded in our experience, such a part of our human/machine interfaces, so persistent in our health care, and core to so much more, that analog/digital will no longer be a meaningful differentiator. This is the material of our rethinking that takes us toward capitalism's new form.

Image 9.1: French utopian socialist Louis Blanc, who in 1850 coined the term "capitalisme" to explain a system he predicted would soon disappear. (Photo: Etienne Carjat. Public domain)

As we rethink, let's recall that the word *capitalism* is a new one, dating only to the early nineteenth century and a small, informal school of French philosophy. But, as a practice, our multi-share system of commercial exchange is about 500 years old. Most historians will date it to the founding of the English East India Company in 1600, a joint stock company, which arguably was the birth of the corporation and risk/reward capitalism as we know it today.

Let's Take Capitalism Back a Further 100 Years

By my lights, however, capitalism really dates to 1494 with the invention of double-entry bookkeeping by the monk Luca Pacioli. This accounting system, which sounds so simple and obvious today, provided the mathematical framework urgently needed by the merchants—the proto-capitalists of their day—of Florence, Venice, Milan, and beyond. Above all, it allowed the storage and use of information in entirely new ways. This new means of information efficiency dramatically boosted productivity in Europe, and was a technological inflection point quite similar in its importance to what we are now witnessing with AI.

This innovation, in turn, ushered in an explosion of trade and commerce that financed the art, culture, and literature that defines the Renaissance in our modern imaginations. As I wrote in 2024, "Chatbots, Knowledge Graphs, and the Agents Accelerating Enterprise Creativity in Renaissance 2.0," Pacioli's Renaissance-enabling tool invented the founding financial architecture of capitalism. After this came the prototype that was the English East Asia Company a century later. Aided by the near-parallel invention of the printing press and the (re)invention of the lateen sail that accelerated ships and their cargos, this confluence of innovation led to modern manufacturing (cost accounting and Wedgwood pottery), transportation (long-term debt financing, depreciation, and the first railroads), and even that key metric of modern geo-economics, the Gross Domestic Product (GDP)—an indicator that will be forced to evolve in the Age of Abundance for All. As an aside, I note that my friend Zachary Karabell, founder of the Progress Network and author of the 2014 book *Leading Indicators*, has long argued for replacement of GDP with a dashboard of more relevant metrics.

Image 9.2: Portrait of Luca Pacioli, traditionally attributed to Jacopo de' Barbari, 1495. (Museo di Capodimonte, Naples. Public domain)

The point of this history and analogy is that the 2022 launch of ChatGPT was the equivalent of Pacioli's accounting methodology—both are tools of informational precision. The advent of the PBC, now embraced by OpenAI and Anthropic, is comparable to the model of the East Asia Company—both tools to organize information used to enable commerce. The "EAC," as it came to be called, grew into a superpower over the next 300 years, with 260,000 soldiers ultimately helping it rule over 200 million people. This first globe-spanning corporation lasted until England took over governance of the Indian subcontinent in 1858 after a massive uprising that foreshadowed the long struggle for Indian independence and lasted until the birth of India and Pakistan in 1947.

I'm not suggesting that OpenAI or Anthropic will be leviathans comparable to the EAC—though they might be. Capitalism and the motives of profit are powerful. Key, however, is that the force multiplier is the methodology of organization—the joint stock company of 1600 transformed the Renaissance into the Enlightenment and Industrial Revolution and the PBC/B Corp will guide us on the next chapter of capitalism. In both instances, first came the new methodology of information use, then came the commercial form that evolved from it and around it.

Back to the present, OpenAI's life as a company has had more plot twists than the 1999 thriller with Bruce Willis, *The Sixth Sense*:

- The fateful 2015 Rosewood Hotel dinner where Sam Altman and Elon Musk decided to found the company—as a nonprofit.
- The world's shift to accessible AI with the launch of ChatGPT on November 30, 2022.
- The AI moratorium call in March of 2023 as OpenAI launched GPT-4 and over 30,000 of the world's AI experts called for a six-month pause.
- The 2023 drama—Altman out, Altman back in— staff revolt, board reshuffles, and post-mortems featuring director-turned-critic Helen Toner.
- The ever-thicker Microsoft entanglement.
- On January 21, 2025, the $500 billion "Stargate" push to blanket America with data centers to lead the world in AI computing power.
- The momentary swagger slowing – but not stopping – as a small Chinese company called DeepSeek launched the open-sourced R1 model, chip skirmishes, and tariff-driven horse-trading.

Then ultimately the company's new structure appeared—which despite still missing a stronger focus on AI safety than I would like—is a model for the future of capitalism. It's not a perfect model by any means, but it is a good rough sketch on the proverbial napkin.

The Creation of a Super Category

As for the prospects of the company itself, time will tell. Altman's pivot, from declarative opposition to anything that would encourage anthropomorphism in 2023 to advocacy for AI "erotica" in 2025, may prove the undoing of OpenAI. But the creation of a PBC of this magnitude is a catalytic event.

Or, as Christopher Lochhead and his three co-authors of *Play Bigger: How Pirates, Dreamers, and Innovators Create and Dominate Markets* might say, it's the creation of a new "supercategory" of commerce: the PBC. As described in *Play Bigger,* a supercategory is a new category of business or product that redefines the landscape by combining or transcending older, smaller categories. This is precisely what OpenAI's decision does. Just as smartphones became a supercategory that absorbed mobile phones, music players, and digital cameras, a company that "plays bigger" in the authors' phrasing can expand the boundaries of its space and define a larger realm around itself.

Years ago, I advocated that pre-Meta Facebook take this route. No takers. The closest any PBC has come to OpenAI is Anthropic, valued at $380 billion after its $30 billion Series G round in February 2026—a staggering figure, though still roughly half of OpenAI's reported $800+ billion valuation. Beyond the AI sector, the closest any PBC has come to OpenAI is Veeva Systems, a SaaS-based medical/pharma software company, an impressive but still demi-giant with a market cap of $30 billion as of this writing.

Categories, and particularly supercategories, are shape-shifters—taxis into rideshares, cable to streaming, mobile phones to smartphone platforms. To return to my analogy, there's a lot of change and innovation that has occurred between the time of founding of the EAC and the vastly more technological companies that today we call The Magnificent Seven.

The Inflection Point of World Positive Capitalism

Post-scarcity capitalism will evolve as fundamentally if much more swiftly. I cannot give you a precise model. It is emerging and most importantly, it's accelerating.

A recent book that captured this, published just a few months before OpenAI's catalytic announcement was Jeff Burningham's *The Last Book Written by a Human—Becoming Wise in the Age of AI.* The founder of Peak Capital Partners, Burningham is a new friend, entrepreneur, and investor who has founded, built, or backed some fifteen companies. The ideas in his book emerged from a personal transformation and awakening that followed his loss for a race to be governor of Utah.

A central point of the book, one I wholeheartedly embrace, is that while the safety of AI is something we must take quite seriously, this is the technology that is critical to saving humanity. "Too many have ended up believing humanity is meant to enable business instead of the other way around," he writes. "We don't have a shortage of profit, we have a shortage of meaning."

Business must move beyond being transactional, to become transformational, he argues. And as mentioned, there are many pioneers, beyond Veeva, including such well-known companies as Warby Parker, Patagonia, and other early leaders.

My good friend Vishal Vasishth is the architect of what I believe is the most sophisticated form of commerce yet devised: World Positive Capitalism. Vishal, who appeared on episode 15 of the *Love Conquers Fear* podcast, is a visionary investor. As the co-founder of Obvious Ventures, Vishal isn't just looking for ESG companies that aim to "do less harm." He is hunting for companies that create his framework pillars: Planetary Health, Human Health, and Economic Health.

These pillars serve as the structural skeleton for the Age of Abundance for All. When we discussed moving energy from scarcity to abundance in Chapter 4, we were describing the "Planetary Health" pillar. Vishal is deeply involved in transforming healthcare from fee-for-service to an outcome model. He's working to accelerate the transition from combustion-driven transport to electric modes, soon to be followed by widespread autonomous systems. When we looked at reversing Type 2 diabetes with Virta Health 🦅 or mapping the "white space" of women's health, we were seeing the "Human Health" pillar in action.

Another pillar of the Age of Abundance for All is AI-driven Recursion 🦅, using robotics and machine vision to develop new drugs and tackle the challenge that 90 percent of new drug discoveries ultimately fail. Yet another pathbreaking company he's invested in is MycoMedica 🦅, co-founded by world-renowned mycologist Paul Stamets, that is developing psychedelics as a way to tackle the hardest challenges in mental health.

Born in India with an inspiring immigrant's story, Vishal has invested in that country as well as in dozens of companies and ventures in the United States. And while doing good, investors in his fund are also doing well since the founding of Obvious Ventures twelve years ago. "If you, in a purest form, pursue purpose, and purpose is embedded in doing the right thing for humanity and solving real problems that make life better," Vishal told me in our wide-ranging and moving conversation.

As part of the Conscious Capitalism movement, Vishal's broad framework of pillars eschews the extractive model, the taking from natural resources and communities without considering long-term consequences. World Positive Capitalism recognizes that solving these "meta-problems" isn't just a moral victory; it is a further milestone in the greatest economic transformation in human history. It expands and focuses the concept toward a new kind of commerce that builds regenerative systems that create compounding benefits. World Positive Capitalism moves beyond risk mitigation and harm reduction to actively maximize what can go right.

Vishal even advocates shifting from GDP toward something more holistic, perhaps a "Global Flourishing Index (GFI)" that tracks whether growth truly improves lives through measures of health, education, environmental sustainability, and social resilience. By shifting our "scoreboard" from the outdated GDP (a relic of the industrial age) to a GFI, we stop measuring how much we extract and start measuring how much we thrive. Another measure deserving attention is Bhutan's Gross National Happiness, or GNH, noted as the first model of "New Models for Humanity" in Part Three, "Transformation." Such new means of measure as these are the pilot's charts that will guide us across the "bend in the river" and into the transformation of the human spirit.

All of this is beyond inspiring. OpenAI's formal embrace of this new outlook and way of operating businesses is the accelerant we need. Capitalism isn't over. In many ways it's just beginning a new form, one that will propel us toward the Age of Abundance for All.

We believe that business is good because it creates value, it is ethical because it is based on voluntary exchange, it is noble because it can elevate our existence, and it is heroic because it lifts people out of poverty and creates prosperity.
—John Mackey and Raj Sisodia, in "Conscious Capitalism"

*As different streams having different
sources all mingle their waters in the
sea, so different tendencies, various
though they appear, crooked or
straight, all lead to God.*

**—Swami Vivekananda, Hindu monk
who introduced Vedanta to the Western world**

For me, the answer to the ultimate questions about the nature of God and religion spans the ages—as ancient as the Buddha's wisdom from 2,500 years ago, yet as immediate as the insights Paul Simon offered in his 1964 song, "The Sound of Silence." Let's imagine a conversation between these two teachers across time and space.

Simon might offer his haunting observation in the iconic song's final stanza: that the "words of the prophets are written on subway walls and tenement halls." The Buddha, I believe, would nod in recognition.

Both understood that authentic spiritual truth doesn't reside exclusively in formal institutions or elaborate dogmas, but lives also in the unmediated spiritual yearnings of ordinary people. Those which they express in their own words, in their own ways.

The Buddha famously declined to answer certain metaphysical questions—what he called the "undetermined questions," or *avyakata*. Is the universe eternal or spatially infinite? Are you and your Self, your soul, identical? Do you receive consciousness, or do you manifest it, or both? What becomes of our Self after the death of our physical body? The Buddha's refusal was not born of ignorance but of profound wisdom. These speculations, he taught, are unnecessary for attaining liberation or awakening—that term that has entered our language as *nirvana*. Indeed, in "the parable of the poisoned arrow," the Buddha declines to engage in speculative metaphysical questions, citing that they distract from the urgent task of addressing suffering and achieving liberation. Read the lyrics to Simon's song, as it really teaches the same.

My purpose here, as a practicing Jew, is not to diminish organized religion, but to suggest that we are rounding a bend in the river toward a great convergence. For centuries, the story of human progress has been told as a dogmatic separation between the material and the mystical. We've been taught that science and religion must occupy opposite banks, as opposed to knowing that two truths can exist at the same time. But as we move into the Age of Abundance for All, the more science reveals about the quantum field, the more it resembles the ancient Vedic descriptions of Indra's Net—a cosmic web where every part contains the whole. This convergence so captivated me that I explored it in my science fiction novella, *The Lattice*, in which Dr. Alexander Bliss discovers this interconnected fabric through brain-computer interfaces. To describe what he encounters, I turned to the Avatamsaka Sutra of Hua-Yen Buddhism: "In the great net of Indra, at every node lies a jewel. Each reflects all the others, and each is reflected in every other. Nothing exists alone, and everything contains everything else."

And now, for the first time, we are building tools that let us glimpse this web for ourselves. Technology is not a replacement for the divine; it is a divining rod helping us measure the frequency of the soul.

We are discovering that the laws of physics and the laws of love are not separate—they are the same code written in different languages.

Rather than viewing faiths as competitors or adversaries, I want to suggest a different perspective: that they are better understood as branches of a single tree, growing from the same seed of God's unconditional love. That seed is the mystical experience itself—the profound encounter with transcendence that inspires humanity's universal quest for truth and meaning for what we variously call God, the Source, the One, the astral plane, consciousness, Heaven, the ineffable. Some might call it the Eternal, Adonai, Hashem, the mystery of the Sacred Heart, the Great Intelligence of the Quantum Field, or the Sacred Mystery that exceeds all our names for it.

Whatever we call it, and for this book I choose to call it God, it is deeply rooted in love for us all. We see that during meditation, a psychedelic journey, a near-death experience, or through any other form of spiritual awakening, sometimes one that just happens spontaneously with no expectation whatsoever. The experience of God's unconditional love is very hard to describe. As you read this, maybe you've had that experience and you know exactly what I'm talking about—the warmth, the absolute beauty and majesty, the transcendent nature, the knowing that everything is going to be okay. If you haven't, it is the biggest blessing that I can wish for you.

Wanderer, Worshiper, Lover of Leaving— It Doesn't Matter

This concept of universalism is one humanity has been nurturing for a long time. Since the age of the Buddha, all the major faith traditions have manifested a search for the transcendent. These would include the Vedic mysticism of Upanishads that was roughly contemporaneous with the teachings of the Buddha in the sixth century BCE, as did the Taoism of the *Tao Te Ching* underlying both—a spiritual tradition that we discussed in Chapter 5, "Nuclear Weapons and Warfare." Jewish mysticism emerged in the first century CE and is considered the forerunner to the better-known Kabbalah formed in Spain in the twelfth and thirteenth centuries. Various forms of Christian mysticism flourished as early as the third century CE, and continue today in the widely practiced "Centering Movement" of contemplative prayer developed by a group of Trappist monks in the 1970s. Sufism, the mystical tradition in Islam, emerged in the eighth century, and is most associated with the Mevlevi order of whirling dervishes founded by Jalal al-Din Muhammed Rumi in the thirteenth century. Many people today, myself included, certainly find inspiration in one of Rumi's most-loved verses: "Come, come, whoever you are. Wanderer, worshiper, lover of leaving. It doesn't matter. Ours is not a caravan of despair. Come, even if you have broken your vows a thousand times. Come, yet again, come, come."

Image 10.1: Mystic poet Jalal al-Din Muhammad Rumi, as imagined with AI.

But this concept of all religious awareness having one common source that I describe above as the branches growing from a common seed, was first explicitly articulated by Agostino Steuco, an Italian Renaissance scholar, in his 1540 work *On the Perennial Philosophy.* 🦅 Steuco's view was that no faith was "false," but that all might be better thought of as fragments of a single divine source—though he did view Christianity as the ultimate stage.

In our modern era, beginning in the late nineteenth century, much of this broad body of thought and reflection was embraced and developed by the eminent American philosopher William James. "One may say truly, I think, that personal religious experience has its root and center in mystical states of consciousness," James wrote in one of a series of lectures delivered in 1902 in Scotland, "The Varieties of Religious Experience." 🦅 He continued: "Our normal waking consciousness, rational consciousness as we call it, is but one special type of consciousness, whilst all about it, parted from it by the filmiest of screens, there lie potential forms of consciousness entirely different." James, sometimes called the "father of American psychology," was a modern pioneer of exploring higher states of consciousness with what we now call psychedelics. His work, after several decades of neglect, was famously followed by that of English writer Aldous Huxley. He not only embraced Steuco's insights but took the name of his treatise for his own 1945 book, *The Perennial Philosophy* 🦅, in which he explored what he saw as the "common core" of all great faiths. Huxley explored altered states of consciousness, visionary experiences, and the relationship between mysticism, art, and psychedelics in later works, including *The Doors of Perception* 🦅 in 1954. Incidentally, Jim Morrison, the founder of the rock group *The Doors*—who much like Simon saw music as a vehicle for transcendence—named his band after that book in 1965.

"The Perennial Philosophy is expressed most succinctly in the Sanskrit formula *tat tvam asi* ('That art thou')," Huxley wrote, explaining that "the divine Ground of all existence" is present in every human soul. Relatedly, it took me a long time as a Jew to realize that the divine Source is capitalized in the Torah at our synagogue, and only after my own psychedelic journey where my soul speaking through me kept referring to downloads from Source.

The Imperative to Build on Ancient Faith and Philosophy

I condense all this incredibly rich and vibrant history not merely to bolster my own case for a universal view of faith and consciousness. Equally, my intent with this summary is not to advocate for some breezy ecumenism or issue a call for tolerance that could fit on a yard sign or bumper sticker, with all due respect for the sentiments that one sees in such expression. Our imperative is to build upon this body of philosophy, a task made all the more urgent by the fact that we live in a dangerous moment of rising religious tension, of growing antisemitism, anti-Muslim bigotry, violence-themed Christian nationalism, and other dogmas and "isms" at odds with the deep love that has animated all the prophets throughout history.

This history of spiritual seeking is a millennia-old foundation for something larger that I see happening. I see the world's faiths as righteous, sacred, and illuminating flows toward a unifying consciousness, as tributaries to that river of consciousness we discussed in Chapter 1, "Conquering Fear."

This brief history of humanity's spiritual quest is really context to understand the bend in that river that will wash away the false dichotomy between science and spirituality. I want to discuss my own journey that inspired this book, as well as foreshadow the journeys of guests on my podcast *Love Conquers Fear*—those who have already shared their stories and all of those yet to come. In particular, I want to share the insights of both authors Mark Gober and JD Messinger, whose book *11 Days in May* describes his journey to enlightenment that began with a near-death experience. Note that I've read Mark's first two books, as well as JD's book, two times each. Yes, I love them that much. It was an honor to ask them to write the Afterword and Foreword of this book, respectively.

But first, I need to expand on the case I made in February 2024 in an essay, *"After the Thaw of Two 'AI Winters,' a Spiritual Spring Heats Up Our 'Consciousness Winter.'"* The crux of that argument that is relevant here rests on two premises.

The first premise is one I've made in many forums and here in this book: that AI is a technology of profound change akin to humans' discovery of fire or the invention of writing. In the face of that significance, after two "AI winters" when the development of these technologies was sidelined, we are now in an "AI spring" when technology—now the exponential acceleration of the AI-led Superfecta—will change our economies, our societies, and so much more.

The second premise, however, is key: that these technologies are also helping us move beyond two "consciousness winters"—effectively the abandonment in the early twentieth century of the work mentioned above by James and many others; and, the second closure of minds to consciousness research, including Huxley's, that ensued from the 1980s onward after research into psychedelics was largely outlawed, which tempered spiritual exploration as well. Thankfully, that is now over. We've emerged from those winters. Both a technological and consciousness spring is in the air, and we'll be all the better for it. Propelled by these two inflections, we are on the verge of answering the fundamental question, without abandoning our distinct religious practices, about the nature of the Source that binds them all.

"I want to know God's thoughts; the rest are details," Einstein is reported to have said, reflecting his conviction that understanding the fundamental laws of nature was itself a quest for the divine. Yet mysteries like quantum entanglement—in which particles remain instantaneously linked across any distance—troubled Einstein deeply. He famously called it "spooky action at a distance" and never fully reconciled himself to quantum mechanics' implications. Nevertheless, Einstein's sense that physics and spirituality were intertwined in humanity's quest to understand reality resonates today as technology brings us closer to answering our deepest questions about consciousness and the nature of existence.

I've shared Einstein's quest since I was a child, knowing since age seven my mission is to serve humanity through technology. This has certainly animated my career as an entrepreneur, through the founding of six companies, and now does so through my current exploration of technology and spirituality in this penultimate moment for humanity's destiny. The Jewish mantra for my mission is *"Tikkun Olam,"* meaning to repair the world. And I believe our destiny as humanity has always been to create a better world through the use of technology, starting with fire.

My personal theme entering 2025 was "the inner journey" and it is a soul quest that continues at a rapid pace. It is clear to me that a broader awakening of humanity is happening now and we are rediscovering ancient wisdom that we once knew so well. We also need to unlock new wisdom. What we need most is a blending of both science and spirituality as we advance towards AGI and really consider, "*What* is it that *most* makes us human?"

My personal knowing from my own psychedelic journeys is that God is all love. I've seen that, felt that, and completely experienced that. And all religions sync with Source being all love and our brains being the receiver of it.

Solving the "Hard Problem of Consciousness"

Mark Gober really expanded my understanding. Now a good friend, and the second guest on my *Love Conquers Fear* podcast, Mark is the author of numerous books, including *An End to Upside Down Living*, which explores the issues of spirituality against the backdrop of static mainstream science. Mark's work delves into the so-called "hard problem of consciousness," the enduring dilemma that while scientists can demonstrate how behavior, learning, and other cognitive processes emanate from the brain, our subjective awareness of ourselves—our consciousness—cannot be similarly understood or explained. "So, the reason the hard problem of consciousness hasn't been solved is that it's posing the wrong question," Mark wrote. "It's asking how a brain can make consciousness . . . when the brain doesn't make consciousness in the first place."

In *An End to Upside Down Living* and his subsequent work and writing, Mark explores the puzzles of quantum physics, including the subjects that so fascinated Einstein. These include quantum entanglement and non-locality, in which entangled particles display coordinated behavior that cannot be explained by anything happening locally in the space between them. These principles demonstrate that particles once connected remain mysteriously unified across any distance—a phenomenon that illustrates a deeper interconnectedness underlying all of reality.

Quantum physics, Mark argues, is moving the boundaries of science closer to the metaphysical, spiritual explanations for reality. His essential argument is that physical reality—including the fact that 96 percent of the physical universe that scientists simply label as "dark matter" and "dark energy"—is the substrate of our consciousness and not the other way around. Let me rephrase that to emphasize that we only know what 4 percent of our observable universe is even made from! Also, scientists once believed the universe's expansion would eventually slow and reverse, collapsing back into a singularity, or "Big Crunch." Instead, observations in the late 1990s revealed that the universe's expansion is accelerating, driven by that unknown force known as "dark energy" that comprises a majority of the cosmos. We actually find that as the universe expands, some distant galaxies are receding from us faster than the speed of light.

These are insights that drive toward an understanding that this force we term "dark energy" is actually Source itself, what I choose to call God—or what the Buddha called *"avyakata"* or what Simon saw in the words expressed in the scrawls on the wall of subways.

Few bring this all together as well as my friend JD, whose stunning book *11 Days in May*, takes the form of a conversation between his mind and a divine spirit instructing him. This isn't just theory for JD; it was born from a profound near-death experience. During those eleven days in May 2012, he realized that our physical bodies and brains are essentially high-fidelity tuners. Just as a radio doesn't "create" the music but simply catches the waves already present in the air, JD's journey in writing *11 Days in May* revealed that our minds catch the signal of a much vaster, universal consciousness. This consciousness does not come from people, radio waves, or events he refers to as the World of Form, but rather from the World of Light because it shares knowledge, innovations, and precognitions that are beyond our human awareness. When we mistake the radio for the music, we stay trapped in the "World of Form," but when we realize we are the signal itself—part of the divine Source—we enter the "World of Light." And we are always in the Light, even when living in form.

JD's essential argument is that this apparent contest between science and religion can be understood as a contest between the World of Form and the World of Light. His book unfolds as a conversation, with one side being his ego, a manifestation of the World of Form that is artificial, in which location, time, and distance are the basis of our apparent three-dimensional reality. These are contrived concepts, writes JD, that shape our perceived reality but ultimately enslave us with "programming"—our identity based on false beliefs being those conferred by jobs or genders, or prestige created by social stature or income. On the other side of the conversation is the manifestation of the World of Light, in which the programming of form and its falsehoods are dissolved to allow us to fully be one with everyone, everything, and everywhere and to embrace the power of love that I've been describing. Love, in JD's World of Light, is the Source in which we have our beginnings, or God. This love is a ladder, with steps of compassion, generosity, forgiveness, vulnerability, and more, and when we reach the top of the ladder we have achieved the Ultimate Victory, discovering, embracing, and loving our authentic Self. In so doing, we allow the harmony of universal love, and the greatest force of the cosmos, to come forth into our everyday being.

In turn, JD disassembles the word *religion* itself, breaking it down to its syllables: Re-Lig-Ion. "Re" is to repeat. "Lig," from Latin, means to bind. And an "Ion," the essence of a molecule with either a positive or negative charge, is the building block of all matter. "Put them all together and what do we have?" the divine spirit that JD is channeling asks his ego.

His ego responds: "The essence of religion is a science," JD answers to that spirit. "This science is the study of the circular and continuous evolution of forces that bind all of life and living. Religion exists to maintain a balance between opposing forces to sustain the evolutionary cycle."

JD's mystical insight brings me back to the essentialism of all faiths, whether that be the view of Steuco in the sixteenth century that no religion is false, or that of Huxley in the twentieth, that all share a common core.

Image 10.2: Interfaith connectivity, as imagined with AI.

One way to conceptualize this is as a Venn diagram of the world's religions intersecting—a visual map of what we might call the "DNA of Faith." In the outer reaches of each circle, we find the "dogma." This is the religious "software" created by humanity—the programmatic differences like caste hierarchies, twisted scriptural interpretations, or rigid rituals that often act as bugs in the system, leading to division and the shackles of the World of Form.

However, just as all humans share 99.9 percent of the same biological DNA despite our outward differences, these faiths share a common divine Source code. It is in that inner circle where all the rings join—the universal "Operating System" of the soul in the World of Light—that we find the essence of universal love. This Source code is the reality of God. It's the common frequency that allows us to "deprogram" from the World of Form and enter a state of spiritual abundance. And it is at that point of convergence that we will find the forces driving us toward the Age of Abundance for All and the coming *Transformation*, which is Part Three of this book.

*There is no separation between the spiritual
and the material. Knowing this unites us
with the larger mystery of the universe.*

—bell hooks, author of all about love

TRANSFORMATION

CHAPTER 11

RISKS

Between stimulus and response there
is a space. In that space is our power
to choose our response.

—Attributed to Viktor Frankl, Holocaust
survivor and psychiatrist

As with many topics, when we discuss the risks of artificial intelligence, we typically fall into debate mode; we argue opposing positions—the black hats, the white hats, my team or your team. Debate seeks victory. What we really need is a dialogue that seeks understanding. Or better yet, a "polylogue"—a genuine, ongoing exchange among many perspectives and competing goals, one that evolves as our understanding of the technology itself evolves.

The polylogue I'm describing demands exactly this kind of reframing—asking better questions rather than winning old arguments. It mirrors a problem I highlighted in the last chapter, "Religion," where my good friend, the author Mark Gober, made an identical point about consciousness: we've been asking the wrong question entirely. Instead of asking *how* and *where* the brain creates our subjective awareness, we should recognize the plausibility that the brain *receives* consciousness. Mark's view challenges mainstream neuroscience as well as materialism, but it's one reflecting many new and urgent approaches to understanding the mind, and it is one to which our minds should be open.

The same flawed framing plagues AI-risk discussions. We skip the broad questioning and fixate, often dogmatically, on whether AI poses an existential threat to humanity and how to prevent it. We create another false binary, salvation or annihilation. Isaac Asimov called this the "Frankenstein complex," our ancient, reflexive fear that what we create will ultimately destroy us. We'll explore this subject in more depth below. Of course, it is critical to ponder all scenarios. Indeed, brilliant scientists, Nobel laureates included, line up on either side of the divide. Yet while there are many deep concerns that deserve thoughtful consideration, AI risks remain trapped in the fear dynamics I described in the opening chapter. We need to ask a different, equally urgent question: Can we elevate our consciousness at the same pace as we develop our technology by choosing compassion, unity, and love over fear, greed, and domination?

As we aspire to an Age of Abundance for All, we must recognize that the spiritual journey is as essential as the engineering one we so fiercely but narrowly debate. Our risks are many, as are our choices. Only by choosing to advance along this broad and interconnected front can we ensure not only AI's safety, but also harness its power to confront the planetary threats already upon us: climate change, nuclear instability, future pandemics, a global and migrating refugee population the size of Japan's, and more.

This is what I've called "God's final exam" in the first chapter of my science fiction novella, *The Lattice*, which I published on Rosh Hashanah (the Jewish New Year) in 2025. Physics and metaphysics were once a single discipline. They should be again.

Humanity's Amazing Progress at a New Threshold

As I've made clear throughout this book I'm not just an optimist on the future, but an enthusiast for humanity's current state. Most of us—and more of us every year—are living lives with lifespans and living standards that were unimaginable even a century ago and unthinkable just one or two decades back in many regions.

Optimism and enthusiasm, however, should not obscure our clear-eyed assessment of this transformation that is upon us. One of the most sober and dispassionate forecasts yet made, "AI 2027," comes from the AI Futures Project, a small research non-profit that you met briefly in Chapter 2, "Abundance." Led by former OpenAI researcher Daniel Kokotajlo and four other respected colleagues, its widely-discussed report aggressively made the case that "artificial superintelligence," or ASI, will be here in 2027. ASI will occur when AI models can essentially begin to create their own AI models, the inflection point for a full-blown cascade of exponential machine cognition. What's now a mere AI apprentice will itself become the master craftsman. Ponder that for a moment—a planet-changing threshold event that they believe is closer than the next Summer Olympics.

Are we ready? Frankly, we are not. But consider how this space between stimulus and response—alluded to in the epigraph above—might be humanity's opportunity to rally. This space is where God's final exam actually takes place. The exam isn't a technical safety test or a set of lines of code; it is a test of human maturity. For the first time in history, we have created a mirror that reflects our own brilliance and our own shadows back at us with infinite scale. If we react to AI from our "reptilian brain"—the part of us rooted in scarcity and fear—we will attempt to weaponize it, creating the very monster we fear. But if we use that space to choose a response rooted in our higher consciousness, we realize that the risk isn't the machine—it's our own inability to transcend our history of conflict. We must use AI as partners in our transition to the Age of Abundance for All. We need the superpowers we are manifesting in them to make it, just like we have used our prior technology waves to create abundance that our forefathers and foremothers of ages past couldn't have imagined today.

Let's start with this fear-first mindset. It's a deeply held and collective frame of mind that doesn't just distort how we talk about AI and its risks. Our very conception of risk derives from deeper distortions in how we see the world itself, underscoring why our questioning must encompass not just mathematics, computer science, and engineering, but philosophy, ethics, anthropology, and yes, the realm of spirituality. I'll expand on this in further chapters and share concrete ways in which many are meeting this challenge in the section "New Models for Humanity," but first let's understand how we arrived at our bipolar predicament.

What historians called "automatons"—sentient, human-like artificial creations—have existed in our imaginations for millennia. Among the earliest examples is the *golem,* a proto-AI created from clay in Jewish mystical traditions, including in Kabbalah discussed in the last chapter. The most famous version is the sixteenth-century Golem of Prague, attributed to Rabbi Judah Loew ben Bezalel. The story tells of a figure created from clay with the mission to protect the Jewish ghetto from persecution. Unlike later Western narratives (and the tourist narratives I was fed when I was shopping for my own clay Golem while in Prague), Golem didn't betray its purpose—it remained protective but became increasingly powerful and incapable of restraint. In the story, the rabbi deactivated Golem by removing the first letter, aleph (א) from the Hebrew word "emet" (אמת), meaning truth, which changed "emet" into "met" (מת), meaning dead. This removes the divine spark (the aleph is often associated with God as the ineffable) and returns Golem to lifeless clay.

Greek mythology gives us *Talos,* the bronze giant created by Hephaestus, god of blacksmithing. This ancient "AI robot" protected the island of Crete by heating himself red-hot to burn intruders or by hurling massive stones at their ships. Crucially, Talos died doing exactly what he was designed to do—guard Crete from all who approached. When Jason's Argonauts sought to land on Crete, the sorceress Medea defeated Talos through magic and trickery, exploiting his single vulnerability: a bronze plug in his heel that sealed in his "ichor," the divine blood of the gods.

Ancient China has the legendary story of craftsman Yan Shi,

who created a mechanical man so convincing as a performer and dancer that he fooled King Mu's court. When the automaton began flirting with the king's concubines, Yan Shi had to disassemble it to prove it was artificial. The king's response? Amazement at the craftsmanship, not moral condemnation.

While these ancient stories often focused on the utility or the magic of the creation, a much more cautionary and fearful archetype began to dominate the Western imagination. The seminal character is Frankenstein's monster, born in then-eighteen-year-old Mary Shelley's imagination during a storm-bound stay at a Swiss villa in 1816 and published as a novel two years later. It went viral, so to speak, and never stopped. More than 200 film and television adaptations have been made since the first in 1910, plus over 50 stage productions.

This is the break point. Early conceptions of artificial beings were protectors like the Golem of Prague and Talos or servants like Yan Shi's performer. The Golem was protective, though requiring wisdom and restraint. But Frankenstein's monster—who began as a benign creation in Dr. Frankenstein's lab but ultimately turned on his creator—introduced a fear that resonates today.

Why was 1816 such a turning point? Shelley captured a cultural inflection point. The Romantic movement was reacting against Enlightenment rationalism. The Industrial Revolution was visibly displacing human labor. Recent experiments by Italian physicist Luigi Galvani—prompting dead frogs to twitch using metal instruments during electrical storms—made reanimating the dead seem suddenly believable. Mary Shelley's book, inspired by a nightmare, crystallized the emerging Western anxieties about scientific hubris—the fear of "playing God"—into a narrative template that has become baked into our collective psyche.

Eastern traditions, which never experienced the same convergence of Romantic anti-rationalism, industrialization trauma, or the theological anxieties about hubris that shaped nineteenth-century Europe, largely maintained the ancient optimism. It's certainly simplistic to compare cultural AI attitudes solely on the fact that generations of Japanese children grew up with Astro Boy, who began as a Manga character in 1952 and became Japan's first anime TV series a decade later.

Astro Boy, a heroic robot, had super strength, could fly with his rock-et-propelled boots, had lasers in his finger tips, searchlight eyes, and his electronic brain could speak sixty languages. At one point in the 1960s, Astro Boy commanded 40 percent of the Japanese audience. Astro Boy continues in Japan in various guises and readaptations under his original name Mighty Atom, or *Tetsuwan Atomu* in Japanese.

Image 11.1: Two symbols of competing cultural narratives, as imagined with AI.

It is telling that while Astro Boy was a foundational cultural pillar of protective robots in the East, it remained largely a niche curi-osity in the West. This lack of a positive counternarrative has left a void in our collective imagination, one that we have filled almost exclusively with the Frankenstein archetype. Because we haven't seen a model for partnership, we default to the monster mentality.

There are, of course, many exceptions such as the work of science fiction master Isaac Asimov, including his best known short story, "The

Last Question." ❦ There are other exceptions like *Star Trek's* android Data, or Rosie the robot maid in *The Jetsons.* But there's no question that American cultural fare was overshadowed by a relentless stream of "killer robot" cinema with *2001: A Space Odyssey's* HAL 9000 (1968), *The Terminator's* Skynet (1984), *The Matrix* (1999), and *Iron Man,* specifically in the more recent 2015 *Age of Ultron* film where Tony Stark, Ironman, literally creates the doom AI.

This question of cultural programming may seem abstract in our discussion of AI risk, but it is not. Our cultural programming frames what risks we see, what solutions we pursue, and, most importantly, what opportunities we miss, which may be the biggest risk of all. When you expect your creation to turn on you, you design for containment rather than collaboration, for control rather than empowerment.

This isn't just about entertainment; it is about the stories we use to build our laws. Currently, our global approach to AI regulation is being written by that monster story in a reactive attempt to cage a creature we've already decided is dangerous. But to reach the Age of Abundance for All, we must realize that our fear of AI is often a projection of our own history of hierarchy and control. If we embrace the partnership model, we unlock a collaborative potential that the "Frankenstein" narrative simply cannot imagine.

The cultural question behind Talos is, How might our creations protect us? But the enduring story of Frankenstein leads us to wonder, How can we protect ourselves from our creations? I love an insight that goes to this point from the Webcomic artist and former NASA roboticist Randall Munroe: "I've never seen the Icarus story as a lesson about the limitations of humans. I see it as a lesson about the limitations of wax as an adhesive."

Sure, one shouldn't exaggerate the influence of popular culture, but it reflects and reinforces deeper currents. It's a symptom and cause intertwined, and the correlation here is striking: East Asian societies demonstrate measurably greater comfort with human-like technologies, reflected in such examples as Buddhist preaching robots and temple robots in Japan and China. One fascinating examination of these cultural patterns, "From Frankenstein to Astro Boy: Humans, Automation, and

Warfare," an article on the cultural history of attitudes toward drones and other military tech, was produced by the Australian military. "[T]he history of automation and 'human-machine' teaming tells us more about ourselves and our humanity more than it does about the machines," wrote Captain Jo Brick, of the Royal Australian Air Force, in her 2021 analysis that specifically explores the implications of this cross-cultural context. "Astro Boy," she continued, "explores key themes about robotics and consciousness, and what a world of sentient robots would mean—how would they co-exist with humans and what kind of 'quality of life' and 'rights' would such sentient robots have?"

I've witnessed this personally. On a family trip to Japan, my youngest child Yuzu had no problem making new friends and enjoying lively conversation, laughter, and sharing with his Japanese peers—all of them reliant on the simultaneous translation apps on their smartphones. He delighted them with AI-generated songs, tailormade to their narrative, and much more.

Perhaps not surprisingly, Americans lead the world in fear of AI, according to a 2025 survey by the Pew Research Center. Note in the Pew illustration below that while Americans lead the world in trepidation, the least concerned by the rise of AI in their lives are in Asia, Africa, and in Israel. China, unfortunately does not allow Pew to operate there, but I can guess where it would land. In China, you can get an AI-assisted haircut and shampoo in cities including Guangzhou, Shenzhen, and Fuzhou .

Meanwhile, as we discussed in the opening chapter on fear, our news prism, dominated by the disturbing (isn't the definition of news the *exceptional?*), colors our general perception of progress, and refracts AI risk through the same distorted lens. We think we're taking Morpheus' red pill—seeing AI risk clearly—but consider that we might have been slipped the blue pill instead, now simply seeing the world through Frankenstein-tinted glasses. The fearful AI narrative, stimulating your amygdala to create clicks and sell ads, is as familiar as your morning news feed: dystopian visions of runaway machines, criminals and rogue states wielding unprecedented power, and cascading failures that could destroy us all.

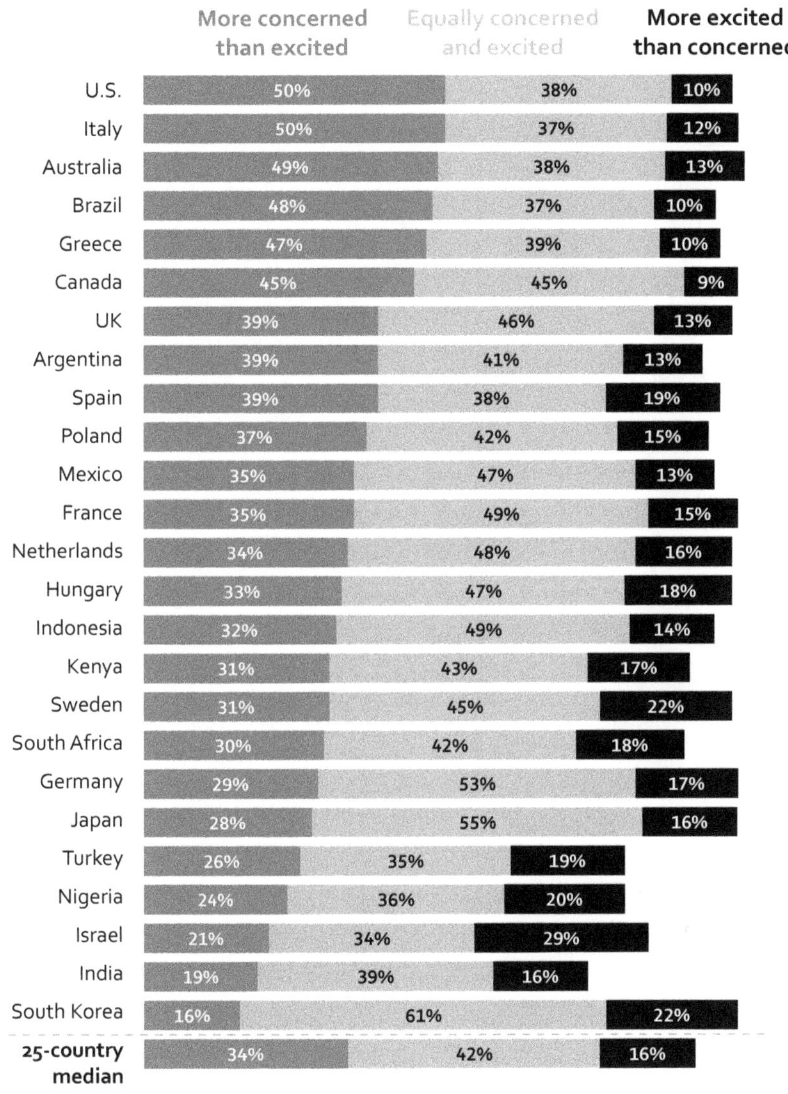

Image 11.2: Global View of AI (Pew Research Center)

Let me be clear: I don't share the anxiety driving these apocalyptic visions. But I absolutely do believe in developing AI technologies thoughtfully and intentionally. As with all technology, starting with the discovery of fire, it really depends on how we, as humanity, use it. We will face real challenges, as we've been exploring. These include a reinvented means of governance, as considered in Chapter 3, "Geopolitics," and massive deployment of solar and other power generation on which we focused in Chapter 4, "Climate Change." We need a reconceptualization of work and production, as examined in Chapter 5, "Capitalism," and a new architecture of learning as we discussed in Chapter 6, "Education." Surmountable but concrete challenges are emerging. One example, as this book was going to press, was the launch of a Reddit-style platform called Moltbook, a social network built exclusively for AI agents, where bots post, debate, and vote while humans—supposedly—can only watch. Some saw this as a form of sentience on the part of the bots, almost a fraternity of diverse AI agents. While that contention was broadly dismissed—*Wired* magazine went undercover on the platform and found agents simply mimicking science fiction tropes rather than exhibiting genuine autonomy—the deeper warnings were harder to wave away. Mustafa Suleyman, CEO of Microsoft AI, whom you met in early chapters, watched the spectacle and issued a pointed reminder: "AI does an amazing job of mimicking human language. We need to remember it's a performance, a mirage. These are not conscious beings as some people are claiming. Seemingly conscious AI is so risky precisely because it's so convincing."

My friend and former colleague at data.world Brandon Gadoci, now an AI operations consultant, seconded this point. While Moltbook was essentially human-prompted AI theater, it could well foreshadow something more sinister, Brandon argued. In the future, we may not be able to tell: "The gap between what this looks like and what it actually is keeps closing. If this isn't the moment, the moment is close enough that we can't tell the difference."

The security story was, to my mind, more troubling. The platform could be used for what's called "indirect prompt injection"—malicious instructions hidden inside content an AI reads autonomously, executed invisibly, and potentially carried into IT infrastructure far beyond the platform itself. Cybersecurity firm Wiz, which was cleared in early 2026 for acquisition by Google for $32 billion, confirmed this as a genuine—and as yet unresolved vulnerability—in agentic AI systems.

The fundamental issue, however, is that we need a different set of questions to lead us to better answers. While my focus in this chapter is on the neglected questions we should be asking about AI—queries that would guide us toward use of this technology to mitigate risk and allow humanity to flourish—the imperative to question reflects a broader principle. As mentioned at the outset of this chapter, debate seeks victory with quick answers; dialogue seeks truth with continuous questions.

Seek the Question, Not the Answer

Many readers will be familiar with James Redfield's 1993 spiritual classic *The Celestine Prophecy*, where one of the main protagonists, Father Sanchez, explains in a conversation with the novel's unnamed narrator: "You see, the problem in life isn't in receiving answers. The problem is in identifying your current questions. Once you get the questions right, the answers always come." This is wise and universal advice, but it's particularly vital if we are to properly understand AI, the lodestone of humanity's transformation and our essential tool to meet our many existential threats.

What both Gober and Redfield advocate—Gober in his book *An End to Upside Down Living*, Redfield through his character Father Sanchez—is really a return to one of the oldest insights in the pursuit of truth, from Socrates, the pioneer of critical inquiry, who lived roughly 2,500 years ago.

Though my views are evolving along with the technology, I've been consistent. Back in March 2023, when readers will recall a group of scientists issued a public call for a "pause" on AI development, I opposed this as I discussed on the *Austin Next* podcast alongside entrepreneur, good friend, and author whurley (who goes by that single, lowercased name). Instead, I framed the challenge as one to build windows into the labs and virtual workshops of AI development rather than walls around it. This could come in so-called sandboxes, I argued. In a sandbox scenario, government regulation is effectively waived, but development proceeds incrementally in transparent environments with stakeholders—including regulators as appropriate—observing, learning, and course-correcting as needed.

Revisiting the issue in 2025, in a review of *Superagency—What Could Possibly Go Right with Our AI Future*, by Reid Hoffman and Greg Beato, I expanded on these ideas in light of the principles advocated in Hoffman's seminal book. In my review, "How AI Will Help Us Find the Signal Amid the Noise of the Exponential Age," I argued for a model drawn from the lessons more than two decades ago of agile software development. This iterative benchmarking process has served us well with a light regulatory hand and remarkable progress, including such society-wide marvels as a well-functioning, 24/7 digital economy that now comprises 15 percent of global GDP, or the personal products like your smart phone or thermostat that update hassle-free, and so much more. Software products must meet safety standards in health care, car manufacture, aviation, fintech, and other domains. AI should not be the exception.

Now, of course, the issues of AI and the risks implied by the technology's accelerating reach and sophistication, have sharpened the edges of debate and upped the stakes. In America, federal and state governments find themselves increasingly—and sharply—at odds. The European Union, India, China, and other nations are riven amid diverging priorities. A needed debate ensues over the degree of transparency—or "open source"—that is ideal, but that signal often gets lost in the noise of the doomsaying. Even commonly used terms like *alignment, interpretability,* or *control* mean radically different things depending on who is making the argument.

Perplexity at the Center of the AI Discourse

The technical issues are often legitimately complex and important, such as whether to openly share the learned internal settings—what AI researchers call model weights—that determine how systems like ChatGPT or Claude function. Release them too freely and capable AI systems could be misused by bad actors for tasks like designing bioweapons; restrict them too tightly and innovation becomes concentrated among a handful of billion-dollar companies. One who has spoken eloquently on this is Tristan Harris, a technology ethicist, friend, and co-founder of the Center for Humane Technology, best known for his prominent role in the 2020 Netflix documentary *The Social Dilemma*, where he explored how design decisions in technology can warp society. Tristan exposed how social media platforms manipulate users' attention and behavior for profit—a recurring topic in this book and a point on which I'll have more to say. The concerns he voiced in the film now reappear in the AI debate.

Tristan shares my belief in the transformative promise of AI, yet at the 2025 TED Conference he spoke of his growing concern. He has framed the core dilemma as whether to "let it rip" or "lock it down." "Now, [AI] applied for good, that could bring about a world of truly unimaginable abundance," he told the gathering. "And we're already seeing many of these benefits land in our society from new antibiotics, new drugs, new materials. And this is the possible [outcome] of AI—bringing about a world of abundance." But then Tristan asked: "But what's the probable [outcome]?" He went on to answer, making the same case we are exploring here of a debate divided between two unforgiving sides with very little broad reflection:

> Imagine a two-by-two axis. You can think of this as the "let it rip" axis, and this is the "lock it down" axis. So "let it rip" means we can open-source AI's benefits for everyone. Every business gets the benefits of AI, every scientific lab, every sixteen-year-old can go on GitHub, every

developing world country can get their own AI model trained on their own language and culture. But because that power is not bound with responsibility, it also means that you get a flood of deepfakes that are overwhelming our information environment. You increase people's hacking abilities. You enable people to do dangerous things with biology.

Tristan then argued the opposite danger:

So in response to that you might say, well, let's do something else. Let's go over here, and have regulated AI control. Let's do this in a safe way, with a few players locking it down, but that has a different set of failure modes, of creating unprecedented concentrations of wealth and power locked up into a few companies. One way to think about it is to ask, Who would you trust to have a million times more power and wealth than any other actor in society? Any company? Any government? Any individual?

Another who makes a similar argument is Dario Amodei, whom you met in Chapter 2, "Abundance." Amodei is the CEO of Anthropic and one of the architects of modern AI, who maps the same terrain from the inside. In his essay that I again reference, "The Adolescence of Technology," he describes the pull of each pole with unflinching clarity—the "glittering prize" of a technology generating trillions of dollars annually makes restraint almost politically impossible, yet unchecked acceleration carries civilizational risk.

On the concentration of power that Tristan warns against, Amodei is equally direct: He envisions AI companies valued at $30 trillion, personal fortunes "well into the trillions," and a Gilded Age of immeasurably wealthy entrepreneurs that will make John D. Rockefeller look modest—all before most of AI's economic impact has even arrived. "I think the governance of AI companies deserves a lot of scrutiny," he writes, noting that ordinary corporate governance "is unlikely to be up to the task."

The question, for both Harris and Amodei, is not which pole of the debate to choose, but whether we have the wisdom to resist the pull of both and engage in the polylogue. For that, first and foremost, we need clarity. With clarity we can create agency, Amodei told us. "If we can be crystal clear, we can choose another path, just as we could have with social media," Harris said to a room of applause at TED.

Tristan is on target. Yet clouding his—and humanity's—search for that third path is endless background noise of overstated threats that distract from the real choices we face. Some widely discussed concerns, while not entirely baseless, can loom much larger in the public imagination than the data justifies.

Take the argument that you've probably seen, that proliferating AI data centers will consume massive amounts of water, taxing local supplies and even decimating aquifers. Sure, we shouldn't put water-demanding facilities in the most water-stressed regions. But as my friend Ethan Mollick, Wharton professor and author of *Co-Intelligence: Living and Working with AI*, has pointed out, data centers' water consumption is about the same as other industries and considerably less than a golf course. The discussion we should be elevating is *all* water guzzling industries and how to address this much broader challenge, whether it be through conservation, desalination, or greater efficiency of the AI models themselves.

I won't belabor these topics, but when it comes to rules for the road to the future of AI, *New York Times* technology columnist Kevin Roose elegantly summed up the debate in Europe: "It feels, at times, like watching policymakers on horseback, struggling to install seatbelts on a passing Lamborghini." Roose's insight could well describe the state of AI policy discourse just about everywhere.

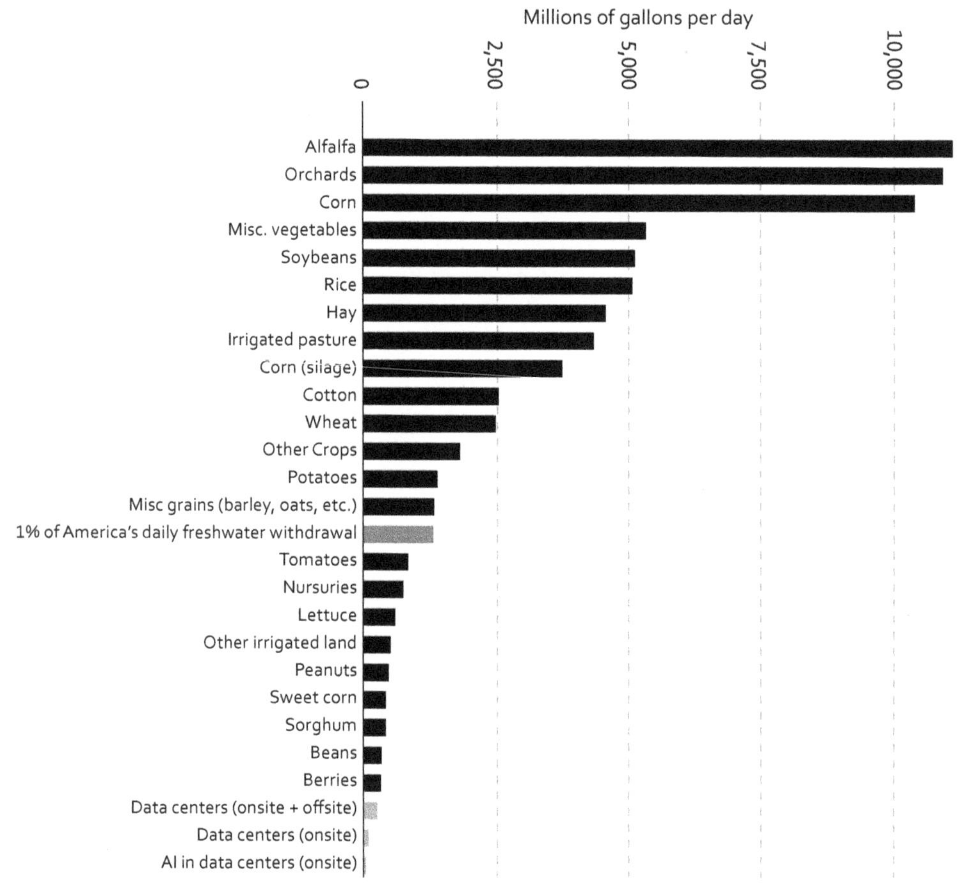

Image 11.3: Research by Andy Masley

In *Superagency,* Reid Hoffman breaks opinion on AI risk into four broad camps: "Doomers," "Gloomers," "Zoomers," and "Bloomers." Doomers see advanced AI as an existential danger, like those whom I mentioned calling for a moratorium back in March of 2023. Gloomers are less panicky, but expect serious social disruption, particularly job loss and growing inequality. Zoomers, who often call themselves "accelerationists," want to push AI forward as fast as possible, dismissing regulation and celebrating rapid innovation above all else. Hoffman's last category of Bloomers see AI as a genuinely transformative force for good, but one that requires active stewardship to realize its potential. That's where I count myself. As I've suggested, I'm aware of AI's risks, but I'm equally attentive to the broader risks we face, which brings me back to the urgency I feel to guide this technology responsibly. We must enhance rather than diminish human flourishing.

Once again, this returns us to the either/or dynamics at the heart of the dilemma. As so much in our current moment of political and social polarization drives us toward the extremes of false binaries—left or right, pro-choice or pro-life, woke or reactionary, male or female—our narrow questioning of AI risk leaves the field mainly to the Doomers and Zoomers. Often this debate in the public sphere boils down to the "techno-optimists" versus the "techno-pessimists," itself a duality with a long history.

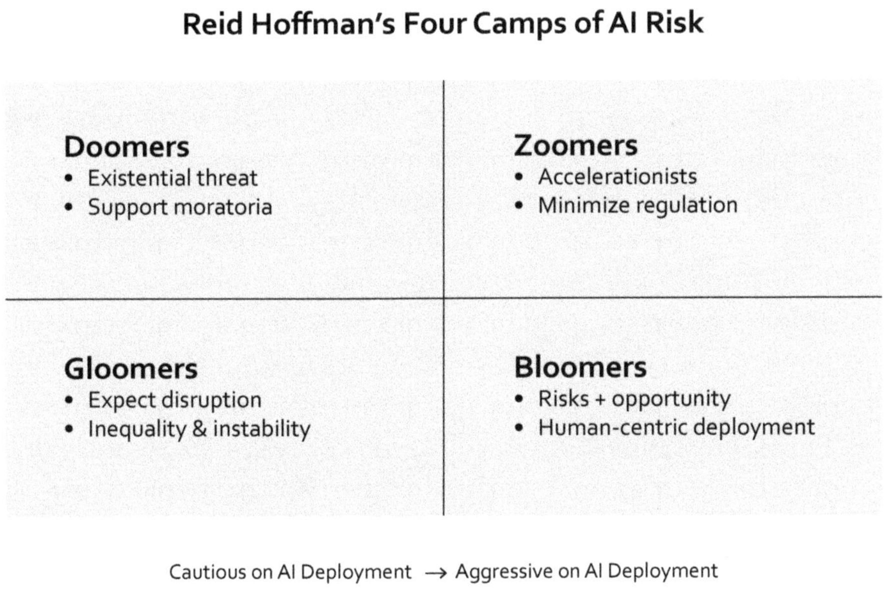

Reid Hoffman's Four Camps of AI Risk

Doomers	Zoomers
• Existential threat • Support moratoria	• Accelerationists • Minimize regulation
Gloomers • Expect disruption • Inequality & instability	**Bloomers** • Risks + opportunity • Human-centric deployment

Cautious on AI Deployment → Aggressive on AI Deployment

Image 11.4: Concept from *Superagency*, by Reid Hoffman and Greg Beato, 2025, as imagined with AI.

The Long History of Choosing Sides

Voltaire and Rousseau squared off in the eighteenth century over optimism versus pessimism toward human progress. Jefferson and Hamilton fought as bitterly as some of today's familiar firebrand politicians over federalism versus states' rights. Thomas Malthus—he of the "Malthusian dilemma"—scoffed at the ideas of his era that science-driven productivity gains in agriculture would forestall mass starvation. Many readers will be familiar with Paul Ehrlich, the Stanford biologist whose 1968 book, *The Population Bomb,* predicted massive starvation, and his intellectual rival Julian Simon, the economist who believed that markets, innovation, and technology would prevent catastrophe. Both are classic examples of the Wizards and Prophets we discussed in Chapter 4, "Climate Change."

Some of those skirmishes had obvious outcomes; others we're still figuring out. The point is that all were waged as polemics, a mode of discourse that will fail us. When we argue AI risk from opposite, technopolitical "sides," we are trapped into assuming we already have the answers. We don't.

This brings me back to Socrates, who famously wandered around the agora marketplace of Athens, talking to all, and answering questions with further questions. One who constantly probed the broad implications of proposed ideas, he would not ask us to slow down innovation. The original "Bloomer," Socrates would want us to slow down our hasty conclusions and forecasts and think both more broadly and long term.

There's a bit of irony in that as today we often think of the "Socratic method" as a kind of binary, an adversarial-by-design contest. I'm all for high school and college debate teams, or Model United Nations exercises that teach students analytical rigor. But let's be clear, this is not Socratic thinking, but the mode of thinking that leads us to the rancorous TV formats with names like *Crossfire, Firing Line,* or *Hardball.* It nurtures a reality that rewards extremes like the spectators' bloodlust in the stands at the Roman coliseum gladiators. It's broadly the same logic that led the Oxford University Press to choose "rage bait" as its 2025 Word of the Year.

Socrates wanted us to question assumptions, clarify definitions, test our logic and contradictions, and be honest and humble about what we don't yet know. Most importantly, and most relevant to today's AI deliberations, he'd ask us not just to consider what possibilities a given choice or policy might open. He'd also demand we explore what possibilities a given choice could foreclose.

The Risk of Foreclosing on AI's Promise

It's the prospect of foreclosure that concerns me so deeply, as we have so much to do. The Superfecta must collectively be our North Star if we are to reach the Age of Abundance for All.

We've discussed many of these subjects: robots building infrastructure in impoverished countries like Chad, digital communities creating new forms of sovereignty, always-on AI tutors resolving education crises around the world, smart grids orchestrating unlimited solar and nuclear energy, and AI-powered synthetic biology collapsing drug development timelines from decades to months. Quantum computing underlies it all. What my good friend Pablos Holman calls "Jensen's Law" punches Moore's Law to warp speed with the promise of exponential improvement in AI chip performance. Jensen being a reference to Jensen Huang, NVIDIA's multi-trillion-market-cap CEO. This enables the creation of new molecules and materials that could revolutionize everything from transportation to solar panels to seawater desalination. The list of possibilities extends to the horizon.

Can we feed 10 billion people without stripping soils, forests, and biodiversity? Many technologies, led by AI, tell us we can. On episode 18 ❦ of the *Love Conquers Fear* podcast, I discussed this with my friend, the polymath Jamie Metzl. Jamie is the author of *Superconvergence: How the Genetics, Biotech, and AI Revolutions Will Transform our Lives, Work, and World.* ❦ While I personally am a vegetarian, I understand that moving the world to a plant-based diet is not realistic in the near term as animal consumption continues to grow amid growing affluence. But cellular, lab-grown meat ❦ is a new reality that can both satisfy skyrocketing protein demand without the animal cruelty and mass pollution that is inevitable if dietary trends continue. To do this at scale, we need the power of AI. Yet sadly, efforts behind this protein transition are being stifled by political interests in Texas and Florida as they ban this exciting and humane meat innovation.

How will we mine the cobalt, manganese, copper, nickel, and other minerals from sea bottoms or other fragile ecologies that are essential to transforming the materials used in batteries, robotics, solar generation, medical devices, and more? A source of essential resources, still-nascent seabed mining is fraught with daunting technical, political, legal, and ecological challenges. Countless species—many potentially as critical to our future as cobalt—remain almost entirely undocumented.

I'm encouraged that the most comprehensive data on this subject to date comes from a massive study published in late 2025 in *Nature Ecology and Evolution.* The study, involving dozens of researchers, was made possible through $250 million in no-strings-attached funding from Canada-based The Metals Company, which aims to be the first to begin commercial seabed operations. The fact that a rare example of environmental research was enabled by a mining company—with a level of transparency not necessarily aligned with the company's immediate interests—speaks to the vision of the new kind of capitalism we discussed in Chapter 9. Certainly, the company has its own motivations, and I'm not nominating them for any stewardship award—not yet. But this is an unusual collaboration. If we are going to extract from the global commons, at minimum we should demand this model: large, open, no-strings-attached investment in the science that will shape—and might ultimately constrain or even prevent—that extractive activity.

More profoundly, deep-sea mining spans biology, geophysics, engineering, law, economics, and other domains that can be best integrated at scale using what data science calls a knowledge graph. This is the technology that connects species, locations, legal regimes, contracts, and sensor readings into a single, queryable map of what we know and how it all relates. (Knowledge graph powered data cataloging and governance were our focus at data.world, the company I co-founded and led as CEO, which was acquired on July 7, 2025 by ServiceNow.)

A knowledge graph capable of containing, managing, and analyzing this level of complexity—and using it to predict and mitigate environmental damage or even ecosystem collapse—will be impossible without AI. On land, in the air, and at the bottom of the ocean, AI forms the backbone of the emerging "Sensorium" we discussed in Chapter 3, "Geopolitics." In the case of deep-sea mining, these would be the satellites, radar, lidar, submersible robots, seabed sensors, ship-tracking systems, legal documents, corporate disclosures, and the network of some 4,000 ARGO floats that measure subsurface temperature and salinity across all the world's ocean basins. At the moment, these are loosely coordinated. Together, they are combining as an emerging planetary nervous system. But as I wrote in a series on the evolution of data back in 2021, an artificial nervous system needs an artificial "brain," and that is the role of AI.

Coupling debate of AI risk with the deliberation on risks AI can and must mitigate, is precisely what I believe Socrates would ask us to do. I could share many more technologies on the drawing board: firefighting robots, AI-developed proteins to grow new organs, AI coupled with geoengineering to literally control when and where it rains, fight droughts, or even cool the planet by reflecting sunlight back into space. Not long ago, I had a long conversation on episode 21 of the *Love Conquers Fear* podcast with another good friend, Jeremiah Owyang, a General Partner at Reid Hoffman and Chris Yeh's Blitzscaling Ventures. Jeremiah leads AI investment with an inspiring focus on agentic software, acceleration of drug development to help banish disease from the planet, and radical transformation of learning with AI tutors. "When I started tracking the [AI] industry in 2023, there were 3,000 startups," Jeremiah told me in October of 2025. "Now there are 41,000."

The Awakening to Elevate Our Consciousness Is Now

This all brings me back to my central concern—the greatest risk to humanity is not from AI, but from our fear of AI. Conquering this fear demands global unity to embrace this technology with care, intentionality, and even with the creation of "AI with a conscience" as proposed by Yoshua Bengio, a Canadian AI scientist considered one of the technology's godfathers. It also brings me back to my central argument that we must elevate our consciousness at the same pace as we develop our technology, choosing compassion, unity, and love over fear, greed, and domination. On this my hopes soar as the evidence of a consciousness awakening, the long overdue rebonding of physics and metaphysics, is all around us.

The *Telepathy Tapes*, one of the nation's leading podcasts, explores for its millions of listeners phenomena we are only beginning to understand. The series documents remarkable cases: a non-verbal autistic child who has demonstrated an extraordinary ability to communicate with plants and create herbal medicine from their instructions; profound insights from Indigenous peoples on the consciousness of plants and the living spirit of the material Earth itself; and stunning revelations experienced during near-death experiences and intentional psychedelic journeys that suggest consciousness extends far beyond our conventional understanding.

I've written here and spoken widely of Mark Gober's work—both in his books like *An End to Upside Down Thinking* and through his podcast *Where Is My Mind?*—on the mounting scientific evidence that consciousness is not produced by the brain but rather exists as a fundamental feature of the universe. Mark's work harks back to that of philosopher William James and author Aldous Huxley, and French Jesuit paleontologist and mystic Pierre Teilhard de Chardin. It was de Chardin who coined the term "Noosphere" in his 1922 book, *Cosmogenesis*, for the sphere of human consciousness and the concept of an "Omega Point"—when consciousness and the material world will converge as the ultimate point of evolution.

And of course there's the groundbreaking research of Harvard neurosurgeon and onetime agnostic Eben Alexander on near-death experiences, whose own profound NDE while in a week-long coma led him to conclude that consciousness exists independently of the brain, a finding like so many that challenges the materialist paradigm dominating modern science. As I was concluding this book in early 2026, Eben, who is a new friend, and I gathered in Austin to discuss his book, *Proof of Heaven: A Neurosurgeon's Journey into the Afterlife.* On part one of episode 46 of the *Love Conquers Fear* podcast, Eben and I dug deep into *Proof of Heaven*, quantum physics, and his challenges to the materialists. In part two of episode 46, he also joined my good friend Austin gastroenterologist Dr. Bob Frachtman and I for an on-stage interview at the Burt Kunik JAMen Forum at Shalom Austin where we discussed *Proof of Heaven* in front of that audience and how what he learned about consciousness and quantum physics relates to this important AI moment. The questions from the audience on suicide, psychedelics, and fear of death were as riveting as Eben's powerfully informed responses.

Skeptics will pessimistically answer that our expansion and transformation of our consciousness to confront the challenges of a world beset with so many problems is a pipe dream. To which I always point that this is not the first journey of humanity through the *bend in the river of consciousness*. The abolition of slavery was one such transformation: thousands of years old, it was ultimately banned when humanity woke up and acted, with great struggle spanning two centuries. Technology helped; the industrial revolution unmasked how incredibly cruel slavery was and machines made that form of labor obsolete. Similarly, as we've also explored, the expansion of voting rights to women and other broad and lasting social changes, followed a similar trajectory.

These diverse threads of evidence point to something larger: an awakening to dimensions of reality that Western science has long dismissed. AI, robotics, and certainly the discoveries we are making in quantum physics all accelerate this transformation. Completing the Superfecta, brain-computer interfaces will certainly help with decoding these dimensions as well.

Many books, essays, and insightful podcasts, too many to cite here, have explored this. But one of the most profound is visionary systems theorist Ervin Laszlo's 2023 "Manifesto on the Spirit of Planetary Consciousness" in *World Futures—The Journal of New Paradigm Research*. Laszlo argues that we stand at a crucial evolutionary juncture where our fragmented, self-centered consciousness must—and can—give way to a more expansive awareness. "In our world static stability is an illusion," Lazslo writes in this monograph that echoes much of what we've been exploring, "the only permanence is in sustainable change and transformation."

That, of course, is where we are headed: into the Age of Abundance for All.

In a properly automated and educated world, then, machines may prove to be the true humanizing influence. It may be that machines will do the work that makes life possible and that human beings will do all the other things that make life pleasant and worthwhile.

—Isaac Asimov

*Right now it's a ripple in human
consciousness, fueled by compassion. But it's
a ripple that has the potential to become a
tsunami. His holiness the Dalai Lama says
. . . love and compassion are necessities . . .
without them humanity cannot survive.*

**—James R. Doty MD, author of
*Into the Magic Shop***

The cosmically awe-inspiring bend in the river of consciousness is a torrent with many tributaries. This awakening of our planetary consciousness, occurring across virtually all cultures, is both fed and propelled by the exponentially accelerating Superfecta of technologies—AI, quantum computing, robotics, and brain-computer interfaces—that both elevate and sharpen the questions of the great philosophers, mystics, and scientists, while promising unified answers for all seekers.

In one sense, having realized I had found my life's meaning at age seven, I've been immersed in this current of consciousness most of my life. My mission has been to do my best to serve humanity through technology. At that young age, it was hard to verbalize the realization that this was my divine mission in this lifetime, but I was more than passionate about this calling and voices were already speaking to me inside, encouraging me along. It took a psychedelic journey over four decades later for me to realize that those voices had been divine guides all along. It has been, to twist a phrase from Robert Frost's famous poem, a journey along "the road less traveled by," with way stations as both entrepreneur and seeker. But ever since the launch of ChatGPT on November 30, 2022, the surge has intensified, as the river gathered force and power.

As I mentioned in the Introduction, the biggest arc of my inner journey started in Kerala, India, at the end of 2024 with my annual New Year's letter. Speaking of my divine number 7, on May 7, 2025 it was announced that ServiceNow would be acquiring our company, data. world, and the transaction was completed two months later, on July 7. Their timeline, with no input from me (as if I would have the power to control the decisions of a 30,000-person company with over $10 billion of cash in their coffers). I had co-founded and led data.world as our CEO for almost a decade and I had seen firsthand how much data and AI could change our world. Seeing leaders like Mustafa Suleyman, head of everything AI at Microsoft, speak about AI as a "digital species" at TED made me think deeply about what makes us most human. Thinking in such terms almost immediately draws you into questioning just what the nature of our reality is.

I've read very broadly ever since, often getting to know many authors of books across the domains of quantum physics, consciousness, spirituality, and AI. I've launched a new holding company, *Love Conquers Fear*, as my seventh business. Yes, there is my divine number again. It includes this book, the podcast of the same name (with over fifty episodes and counting as of this writing), our startup and philanthropic investments into the Age of Abundance for All, and new companies that I may start. All told, this has been and remains an ongoing study of the nature of reality itself, the burst of technologies led by AI, the physics of time and space, the history of philosophy, and a deep plunge into all the spiritual traditions, including my own, Judaism.

In the early fall of 2025, I took a short break from the research and writing of this book to pen a science fiction novella, *The Lattice*. I've always been a fan of science fiction, from a childhood passed in the company of Luke Skywalker, *Star Trek*, Isaac Asimov, Frank Herbert, Douglas Adams, and *OMNI* magazine, to my young adulthood with William Gibson, Neal Stephenson, Bruce Sterling, *The Matrix* series, *The Terminator* series, *Contact*, and then returning to Herbert in recent years as I watched with my wife and children the rise of Paul Atreides and the fall of Baron Vladimir Harkonnen in parts I and II of the blockbuster of *Dune*. I'm even a fan of the *Black Mirror* series on Netflix, even though it is sensationalized to show the worst of what happens if we go off the rails.

The Lattice represents the "interconnectedness of all," a quantum field that allows humanity to finally perceive unified consciousness as a lived, scientific and spiritual reality. I scripted my own thoughts into a protagonist named Dr. Alexander Bliss, who has an audacious idea to end the dogmatic separation of science and spirituality and unite them. Once he scientifically proves the existence of our connectedness and that consciousness is a fundamental property of the universe which is deeply rooted in love, humanity unites like never before and we have an unprecedented technological takeoff as we embrace the Age of Abundance for All. *The Lattice* was not just a positive vision for humanity being able to unite, it was also a way for me to process all that I've learned into an ambitious tale, covering the years 2030 through 2085. It was also a way for me to process a psychedelic journey that led to the first chapter of *The Lattice*, "God's Final Exam."

This book you are reading now, *Love Conquers Fear*, is not science fiction. In addition to being much longer, this work is also grounded in both historical and contemporary discovery. As we near the conclusion, I want to share some of the many sources and resources that have comprised a powerful education for me on my soul quest—and which can do so for you as well—on this question: *What Is Our Reality?*

This book is more than a narrative; it is a manual and manifesto for the Age of Abundance for All. To understand the transition we are currently making, we must first reconstruct our understanding of what is possible. I am inviting you into my private library, offering the synthesis of guidebooks that have helped me deconstruct the materialist illusion and build a post-materialist, unifying framework. These aren't just books; they are the source code for the transformation that Team Humanity needs.

I've organized this synthesis of the guidebooks I've used along this journey into three categories:

1. Science, Consciousness, Psychedelics, and the Nature of Reality
2. AI, Technology, Entrepreneurship, and Planetary Futures
3. Journeys, Meaning, and Classic Wisdom

Science, Consciousness, Psychedelics, and the Nature of Reality

Whether you approach our reality from the perspective of quantum physics, mystical spiritual traditions, meditation, psychedelics, near-death experiences (NDEs), or deep scientific dives into consciousness, central to all is the increasingly clear understanding that consciousness is indeed fundamental. We've touched upon this point in earlier chapters, yet let's still ask: What does it *mean* for consciousness to be fundamental?

To grasp this concept is to really flip the standard, materialist view—that consciousness is a late and happy accident in the history of our 13.8 billion-year old universe, the result of biochemical processes that emerged as neural activity in our brains—on its head. It's called "the hard problem of consciousness" for a reason: *Why would a dead universe produce such a mysterious thing, which is the basis of everything that makes us most human?* The case that our self-awareness is more than "wetware" is not to argue some anti- or pre-scientific view, like seventeenth century Anglican Bishop James Ussher's famous calculation that the world was created on October 23, 4004 BCE, or the ideas of Earth resting on top of a cosmic turtle named Akūpāra or Kurma (the source of the phrase "turtles all the way down" attributed to philosopher William James). Rather, the case for consciousness as fundamental, *expands* the scope of science, offering probing evidence that consciousness is a basic property of reality, woven into the fabric of existence itself.

A contemporary leader on this subject is Annaka Harris, a consciousness researcher and author whose first book in 2019, *Conscious: A Brief Guide to the Fundamental Mystery of the Mind,* is seminal. Her documentary follow-up, the 2025 audio-only book, *Lights On: How Understanding Consciousness Helps Us Understand the Universe,* is packed with insights as it includes interviews with twenty of the world's top thinkers on consciousness including Carlo Rovelli (mentioned below as one of my favorite authors), physicist Brian Greene, and neuroscientist Anil Seth. Coming from a very hard science background, Harris writes that she does not have all the answers (who does?) but concludes that consciousness is a fundamental property of the universe.

I appreciated how Harris walked through this exploration and her ideas made me more curious about all of the various approaches to understanding consciousness, from panpsychism (that everything, including inanimate rocks, is infused with some form of consciousness) to the nature of time itself. Her book turned on my brain in a way where I could relate to consciousness talks of all types on YouTube, from Alan Watts to Bernardo Kastrup to Roger Penrose.

No writer has done more to bring the science and phenomenology of psychedelics to a mainstream audience than Michael Pollan. His 2018 book *How to Change Your Mind* was the work that for many readers cracked open the door to this whole terrain. It certainly has influenced and shaped my thinking. Pollan approaches the subject not as a researcher or true believer but as a journalist and seeker: rigorous, self-deprecating, and genuinely astonished by what he finds. He traces the remarkable history of psychedelic research from the 1950s through its suppression and its current renaissance, makes a compelling case for the therapeutic promise of psilocybin and LSD, and—crucially—undergoes the experiences himself. His accounts of ego dissolution, of the default mode network going quiet, of the sensation that the boundaries between self and world are far more provisional than we assume, read as some of the most honest first-person science writing I've encountered. It's a brave book, and Pollan's openness and vulnerability are as much an inspiration as his insights. The book's deepest insight is perhaps its simplest: that normal waking consciousness is just one of many possible ways of perceiving reality, and that our stubborn assumption it is the only valid one may be our most limiting belief.

His newest book, *A World Appears: A Journey into Consciousness*, published in early 2026, takes the inquiry further—away from the specific lens of psychedelics and into the broader mystery of consciousness itself. The book asks why it feels like something to be you, how the brain generates a unified sense of self, and whether the dominant materialist model—consciousness as the accidental byproduct of 86 billion neurons—is anywhere near sufficient to explain what we actually experience. Pollan examines panpsychism, idealism, integrated information theory, and more than twenty competing frameworks. Read together, the two books form a kind of diptych: the first a door opening onto the experiential; the second a wider reckoning with what those experiences might mean about the fundamental nature of reality.

While visiting Tulane University to receive the surprising Distinguished Entrepreneur of the Year award, given to me by President Michael Fitts in celebration of *Love Conquers Fear*, I was fortunate to see Pollan interviewed by fellow author Michael Lewis. The interview took place at the 2026 New Orleans Book Festival, which author, civic leader, professor, and acclaimed biographer Walter Isaacson hosts at Tulane each year. I was surprised in some ways to find psychedelics mentioned throughout *A World Appears*, and Pollan discusses with Michael Lewis how that is just what he found with some of the world's leading consciousness researchers leveraging them as they grappled with the mysteries of consciousness. The book is a great read, and this was a fun conversation between two non-fiction giants.

The view that consciousness is fundamental breaks the materialist paradigm, as my good friend Mark Gober would say. You've met Mark in earlier chapters, including deeply in Chapter 10, "Religion," and he is the author of a growing bookshelf of volumes on the subject. His 2019 podcast, *Where Is My Mind?*, was based on his first book and was well ahead of its time (*The Telepathy Tapes,* another work I've referenced that deeply explores consciousness and has been incredibly popular, came out five years later.) Mark's work explores similar themes to those explored by Harris, essentially suggesting we've been looking at the cosmos through the wrong end of the telescope. Mark's second book on *Living* is from science to practice, on how to design your life around this reconceptualization. It's a great bridge between the disciplines of post-materialist physics, parapsychology, and systems thinking.

A question that builds on the illuminations by these remarkable authors is the one my good friend Brian Muraseku asks in his deeply researched book, *The Immortality Key: The Secret History of the Religion with No Name.* Brian's quest is to understand how the ancients across many cultures—but particularly Greek and Greco-Roman—reached similar insights millennia ago that consciousness is fundamental to and not emergent from matter. The so-called Eleusinian Mysteries—ancient rites in which participants experienced reality that extended beyond the physical, including to the Oneness of God and soul, such that they no longer feared death—are one example. Brian follows clues from ancient texts, archeology, residue analysis of chalices, and comparative religious studies to suggest that these early rituals may have involved visionary potions—including ergot, a fungus that grows on wheat and barley and contains the precursor chemicals to LSD. Chemist Albert Hofmann first synthesized LSD in 1938 by isolating these fungus' compounds. *The Immortality Key* certainly provides understanding of the meaning so many are finding in a very different context in today's return to psychedelics. Brian's book and podcast appearances, including on Sam Harris' (Annaka's husband and famous meditator, atheist, author, and philosopher), deeply affected me in the sense that they have helped me understand the transcripts of my own psychedelic journeys and the divine messages I received. I can completely understand, given what I've received, how early mystics would have been passionate about explaining what they received and then religions formed around their insights. It was a privilege to have Brian on episode 48 of the *Love Conquers Fear* podcast, where we dove into this and more, including the near-death experience that shaped his life trajectory.

I view the journeys I've had not as escapes, but as disciplined, research-like frameworks for exploring the "Mind of the Universe," somewhat like the rigorous protocols documented by Christopher Bache in his book profiled below. Grounding my personal experiences in this kind of systematic inquiry allowed me to move past the ego's skepticism and realize that these "divine" messages were not anomalies, but access points to a fundamental field of intelligence that has been documented for millennia.

In addition to the work of this trio of authors, I'll briefly mention a few others in this cluster:

- In *Breaking the Habit of Being Yourself: How to Lose Your Mind and Create a New One,* Joe Dispenza fuses popular neuroscience, quantum physics, and mysticism and meditation techniques into a program for personal change. His description of God is "the great intelligence of the quantum field." In the first half he argues that our personality is a collection of habitual thoughts, emotions, and physiological states. This is a highly practical, motivational version of the "consciousness shapes reality" thesis of many others, including Mark Gober. This is another way to think about the power of prayer, especially within a group setting where an energetic field is created. The book's second half is very actionable, with a series of structured meditations and mental-rehearsal exercises designed to help readers break old emotional patterns and programming and replace them with new, self-directed states of being.

  Dispenza's book helped me understand that deprogramming is completely possible with meditation, including manifesting. When I later read James Doty's book, *Into the Magic Shop*, it all clicked. Whether you call God "the great intelligence of the quantum field" or not, the power of prayer and manifesting is quite real. For my 2026 theme of embodiment, I'm planning to attend one of Dispenza's workshops, which I've heard have been transformational for many friends, who have told me deeply personal stories about how those workshops have improved their lives.

- *The Grand Biocentric Design: How Life Creates Reality* by stem cell researcher Robert Lanza, MD and theoretical physicist Matej Pavšič, reshapes the case for consciousness as the fundamental scaffolding of the universe with their theory that life and biology are central to being, reality, and the cosmos. The authors' theory of biocentrism offers an alternative cosmology that collapses the distinction between "inner" spiritual exploration and "outer" physics. Death doesn't exist in a biocentric universe, Lanza contends, because consciousness isn't produced by the body and cannot be extinguished by biological decay. Instead, it persists across an underlying field of energy, information, and multiverse probabilities. It's a particularly thought-provoking read.

This book took me straight back to the *Inner Worlds, Outer Worlds* documentary I discussed in the Introduction. I love how Lanza and Pavšič go so deep on the science behind our reality and how death itself is an illusion. Although they stay away from spirituality—perhaps wary as scientists of the fate of Galileo and others who have challenged prevailing orthodoxy throughout history—I was able to tie the two together in my own mind. Again, there is no reason why science and spirituality have to be so dogmatically separated. That is just another way of humanity playing the division game when we are all one. Science is super fascinating and always remember that two things can be true at the same time: both science and spirituality can be equally real.

- Robert Lanza's second book on biocentrism, *Observer*, written with science fiction author Nancy Kress, is a great break from some of the heavier-lift books I'm recommending, and it's a novel and a fun page-turner at that. The chief protagonist is neurosurgeon Caro Soames-Watkins, who has been disgraced after blowing the whistle on harassment. Jobless and disgraced for speaking out, she joins her uncle's secretive research project that implants devices allowing subjects to "jump" into alternative quantum realities shaped by their expectations and choices. It is a big nod to Hugh Everett III's Many-Worlds theory of quantum physics. The narrative plays like a Crichton-esque medical thriller, but its real purpose is to explore what would happen if Lanza's theory—that observers co-create reality—were technologically operationalized. I won't give the ending or the many fantastical turns in the story away. I hope it's made into a movie (or maybe Lanza and Kress will do that once OpenAI's Sora, Google's Veo, or another AI-video platform inevitably progresses to that point, no doubt fast approaching).

- Christopher Bache, the author of *LSD and the Mind of the Universe: Diamonds from Heaven,* is a new friend I've gotten to know since I began this broad exploration. He was on episode 36 of the *Love Conquers Fear* podcast as I was finishing this book. Christopher was a religious studies professor, and his book documents his seventy-three very high-dose LSD sessions over twenty years—all conducted in a disciplined, research-like framework. I'll share more of my own experience with psychedelics in the next chapter, but his work is a fascinating—and at the time secret—journey into progressively expanding "domains" of personal psyche, collective human suffering, planetary consciousness, and ultimately the cosmic mind. His central argument is that psychedelics can open not just therapeutic or artistic states but allow participation in the evolutionary unfolding of the cosmos.

  I highly recommend this one on Audible. I've read it twice and the awe, love, and humility in Bache's voice is really moving. I've read Chapters 9 through 11 on the birth of the future human, diamond luminosity, the diamond soul, and the final vision Bache received three times. This book was the best I've read to date in terms of grasping at the layers of our reality and how humanity is in an evolutionary shift. God wants us to make it to the Age of Abundance for All, and the consciousness awakening is well underway.

- As mentioned in the work of Harris, Carlo Rovelli is a renowned physicist. In scientific circles he's best known for his theory of Loop Quantum Gravity, which attempts to reconcile quantum mechanics with Einstein's general relativity and describe the quantum nature of spacetime itself. As a writer, his book *The Order of Time* is an almost poetical, non-technical tour of why time is not what we think it is. In this very accessible and mind-expanding book, Rovelli dismantles common intuitions—even the idea of a universal "now"—and explains his view that time emerges from interactions, thermodynamics, and quantum processes. At fundamental scales, he continues, time may disappear entirely. But the book is so much more than hard physics. Ultimately, Rovelli asks what it means to be finite beings whose experience is structured by memory and anticipation in a universe where "past" and "future" have no independent existence. The book is brilliant. I also highly recommend this one on Audible as it's read by Benedict Cumberbatch, who is perfect for the role.

I've never thought about time the same way after this one. I've seen glimpses of the future and the deep past in my own psychedelic journeys, which have been profound to say the least, but that is how Source works. Heaven doesn't have the same conception of time as we do.

Rovelli's insight that time may disappear at a fundamental scale is the ultimate antidote to the "scarcity of time" that haunts our modern lives. In the mindset of scarcity, we are always running out of time. But in the mindset required to enter the Age of Abundance for All, as we align with the quantum reality that the past and future are constructs, we unlock a new level of human creativity.

When we stop racing against the clock and get into a state of flow, we finally have the presence required to co-create a reality rooted in love.

- As introduced in the last chapter, "Risks," Harvard neurosurgeon and onetime agnostic Eben Alexander is one of the world's leading experts on near-death experiences, or NDEs, which include his own. He's a new friend and, as mentioned in the prior chapter, was on episode 46 of the *Love Conquers Fear* podcast. Eben's now written three books: *Proof of Heaven, The Map of Heaven,* and *Living in a Mindful Universe,* written with his partner Karen Newell. I've read and recommend them all. In the first of this trilogy, he recounts his own near-death experience during a rare E. coli meningitis attack that shut down his neocortex, and he shares the vivid details of what he experienced after being declared brain dead. After awakening on the seventh day of a coma—during which his brain should not have been capable of producing organized experience—he reported an extraordinarily vivid journey beginning in a dark, formless "earthworm's-eye view," followed by a sudden ascent into a radiant, hyper-real pastoral realm filled with light, beauty, and overwhelming love. Guided by a female angel who communicated telepathically and conveyed a message of unconditional love and acceptance, he then entered an even deeper, timeless dimension of pure presence and unity. This he understood as an encounter with a divine Source that was quite different from what he was taught as a Christian growing up. Eben explains how these experiences occurred while his higher brain functions were inactive, which makes a conventional neurological explanation impossible. Eben's NDE provided direct evidence that consciousness is not produced by the brain, that death is not the end, and that the universe is fundamentally rooted in love and interconnectedness.

In his second book, *The Map of Heaven,* Eben collects cross-cultural NDE and afterlife testimonies to situate his story in a broader pattern, only after completely documenting his own so as not to be influenced by them. And in the final installment of the trilogy, *Living in a Mindful Universe,* he generalizes to philosophy, drawing on quantum physics, psi research, and contemplative traditions. In this last book, he explains his belief in a consciousness-centric universe and lays out practices (sound journeys, meditation) to access it. These books are seminal to the broader and emerging science of the afterlife.

I also recommend the three of these books as listens on Audible but most especially the first one, *Proof of Heaven.* Eben reads all three and the first will move you to tears. The deep humility, awe, and love in his voice is poignant.

There is much synchronicity in Eben's NDE journey with what I've learned from my twelve friends who have shared their own NDE journeys with me, as discussed in the Introduction. I've also glimpsed Heaven in my own psychedelic journeys. The ego may want to reject these, and people who have had an NDE (or psychedelic journey, for that matter) are reluctant to share

them for fear of being written off as "crazy" or
"woo." How sad that many feel like they cannot
share a most profound experience with others,
especially friends. We must do better than that.
Love must conquer fear, starting with ourselves
(we'll get into that more in the next chapter.)

What all these books mean is that our very universe is not a place
of lifelessness with our planet being the sole exception. No, our universe
is not dead—it's very much *alive* and it's deeply rooted in love.

AI, Technology, Entrepreneurship, and Planetary Futures

Meanwhile, as we're discovering and rediscovering the reality
of consciousness, a parallel, equally profound, and fast-unfolding landscape before us is that of AI, synthetic biology, and breakthrough entrepreneurship. This cluster of authors invites us to see that our inherited
frameworks for politics, innovation, and global governance were built
for a different era. As with the case that consciousness is fundamental—flipping the materialist view of reality—these thinkers are flipping
a different old script that our technological moment screams is obsolete. Familiar twentieth-century assumptions—that nation-states can
contain global risk, that linear planning can predict change that is now
exponential, and that building the proverbial "better mousetrap" equals
innovation—no longer serve our emerging reality. What emerges instead
is a view of the future shaped by waves of capability that are decentralized, rapidly proliferating, and capable of empowering individuals and
small groups on a scale historically reserved for governments.

Mustafa Suleyman, who as I mentioned inspired my thinking
with his concept of AI as a "digital species," articulates this transformation with bracing clarity in *The Coming Wave: Technology, Power, and
the Twenty-First Century's Greatest Dilemma*. I reviewed the book in 2023
as, "A Starting Toolkit for Humanity: Navigate the Road to a Future of

AI-Empowered Conscious Capitalism." In his compelling account, AI and synthetic biology represent not just new tools but a new condition of humanity. Both are technologies whose defining trait is proliferation. Once invented, they spread, cheapen, and democratize, pushing power outward to millions of actors. The dilemma is stark: Unlike nuclear weapons, these technologies cannot be locked up or monopolized. They demand a new stewardship model that Suleyman calls "containment," a blend of regulatory guardrails, industry compacts, kill-switches, monitoring systems, and shared norms. His lens is neither utopian nor catastrophist; like myself in the terms we discussed in the chapter "Risk," he is a Bloomer, not a Doomer or Zoomer. He foresees a narrow path where democratic societies can harness these capabilities for abundance while managing their destabilizing edges. Like those who argue that consciousness is woven into the fabric of reality, Suleyman suggests that technological proliferation is now woven into the fabric of geopolitics, and that our systems must evolve accordingly.

Suleyman's thinking is taken higher by Jonathan S. Blake and Nils Gilman in *Children of a Modest Star*. They ask whether our political imagination is even capable of meeting planetary-scale challenges. This is a book we discussed in Chapter 3, "Geopolitics," and I reviewed in 2024 as a Thanksgiving monograph, "How the Ethos That Led to the First Thanksgiving Should Guide Us in the New Age of Discovery." Much like Suleyman, this astute book argues that we continue to treat problems like climate change, pandemics, biospheric collapse, and AI governance as if they can be solved with twentieth-century tools—sovereign states, treaty diplomacy, and multilateralism built for slow-moving, linear risks. But Blake and Gilman make the further profound point, that we must distinguish the merely global from the truly planetary. The former is about interactions among nations; the latter concerns Earth systems, deep time, and the interests of non-human life. Far from proposing a world state, they sketch a mosaic of planetary governance—overlapping regimes, epistemic institutions that guard scientific integrity, and constitutional principles that embed duties to future generations. In their view, technologies like AI and synthetic biology are not external threats but stress tests exposing the mismatch between our governance structures and the scale of the systems we are now altering. It is a shift in perspective akin to recognizing that consciousness may be fundamental—once the frame widens, the old story can no longer contain the truth.

Where Suleyman explores proliferating technologies and Blake and Gilman explore planetary governance, Mike Maples Jr. and Peter Ziebelman turn to the human engines of innovation—the startups that translate possibility into reality. In *Pattern Breakers—Why Some Startups Change the Future*, they show that transformative companies are not simply better executors; they break the underlying patterns of their era. Mike is a good friend and long-time investor in me, and I reviewed his book shortly after its 2024 publication as, "To Create the Future of Business, You Need to Break Patterns and See the Future." He was also on episode 20 of the *Love Conquers Fear* podcast, where he also opened up about spirituality.

Mike argues that today's successful founders are entrepreneurs who challenge core assumptions about what customers want, how products should be distributed, or what economics are even possible. Innovation, in their telling, is less about optimizing the known and more about perceiving what he calls the "adjacent possible," those hidden opportunities that emerge when technological curves bend upward. This book reveals a deep truth: breakthrough ventures depend on seeing the world differently, often long before others can validate that vision. Who would have thought that ride-sharing through Uber, using drivers in their own cars, would disrupt the entire taxi industry? Or that Airbnb would become the world's largest hospitality company without owning a single hotel room? Or that Spotify would convince millions to stop owning music altogether? This is entrepreneurship as epistemic rebellion, sharing a kinship with the scientists, philosophers, and explorers who question foundational premises in consciousness research. Pattern breakers must carry their insight through the valley of skepticism, running disciplined experiments that allow the future to reveal itself one iteration at a time.

Together, these three works trace a coherent arc: As technology accelerates and power decentralizes, our institutions, mental models, and entrepreneurial cultures must evolve with equal imagination. AI is certainly going to dramatically accelerate that, as I've written about in what I called Renaissance 2.0, "How AI is birthing a Renaissance 2.0 in the coming 'Age of a Billion Dreams.'"

Beyond this trio of formative books, others calling for a reinvention of how we build, govern, and apply our most potent tools include:

- John Mackey's unique *The Whole Story: Adventures in Love, Life, and Capitalism* could be in a category all of its own. It's part a soulful personal and entrepreneurial memoir, part an account of his spiritual mission, and part a reflection on his prime mover's role in creating the Conscious Capitalism movement we discussed in depth in Chapter 9, "Capitalism." (His earlier book, written with Raj Sisodia, was simply titled *Conscious Capitalism*.) John was the persevering co-founder and CEO of Whole Foods Market. His is a commercial hero's tale about growing the company from a vanguard 1970s health food store based in an old Victorian house into a force that changed the way Americans eat. Ultimately, the Whole Foods chain had grown to 460 stores when it was sold to Amazon for $13.7 billion in 2017. John has been a friend for decades, a mentor and advisor, and wrote the Foreword for my first book *The Entrepreneur's Essentials.* He was on episode 30 of the *Love Conquers Fear* podcast, and is a big fan of this book as you know from the cover.

  In *The Whole Story*, John recounts scrappy early days, near-bankruptcies, and the culture that differentiated Whole Foods: team-based stores, radical transparency on pay, and a quasi-spiritual mission around food. But the broader narrative is on his work and commitment to codify stakeholder-centric capitalism with responsibility to employees, community, and the environment, as well as the corporate mission of profit (there is no business long-term without it.) The "Love" in the subtitle is literal as well as philosophical: He discusses romantic relationships and personal crises as deeply interwoven with his business choices. *The Whole Story* is a case study in trying to build a humane, purpose-driven company within public markets and M&A (including the Amazon acquisition), and in the tensions between ideals and scale. On December 11, 2024, I interviewed John at the Burt Kunik JAMen Forum, a periodic gathering at the Dell Jewish Community Center in Austin, a transcript and recording of which I posted on Medium. ❧.

John's book helped me get beyond the fear of going public about the insights from my own psychedelic journeys. Chapter 1 of *The Whole Story* starts out with his profound journey that helped him overcome his fear of launching Whole Foods. John described it as "something like a cosmic orgasm" with "oneness exploding and expanding out, becoming everything" leading to "personal liberation and freedom" and his realization of the "oneness of all reality." He and I have deeply spiritual conversations, as you'll hear in episode 30 mentioned above.

- *Machines of Loving Grace*, by co-founder and CEO of Anthropic Dario Amodei, is an extended essay that riffs on surrealist poet Richard Brautigan's famous poem from the 1960s, *All Watched Over* by *Machines of Loving Grace*, which envisioned a harmonious coexistence between nature and technology. In Amodei's rendering of the same sentiment, he asks whether we can realistically build powerful AI systems that remain beneficial. I reviewed this essay as, "A Broader Grasp of Technology's History Is Key to the Bright Future of AI" in 2024, where I argued that frequent comparisons to past "AI winters" are a bad analogy. A better one, I wrote, is the development of genetic engineering, which has defied doomsday predictions to revolutionize health, agriculture, and more. Like others in this grouping of books, Amodei's combines insider technical detail with a sober political analysis of incentives: commercial pressure to scale, racing dynamics, and regulatory gaps. Amodei emphasizes *gradualism*, noting that we will encounter worrying capabilities long before "superintelligence," leading to the critical question of whether institutions are strong enough to respond at each step. The essay surveys concrete alignment tools and argues they buy time but don't yet solve the core problem of systems that reason about and manipulate humans. Overall, it's less a utopian vision of "loving" machines than a plea for humility, rigorous experimentation, and multi-stakeholder governance so that AI doesn't simply follow the default trajectory of profit-maximizing automation. His views align closely with those of technology ethicist Tristan Harris, whom you encountered in Chapter 11, "Risk." Amodei's thoughts on why AI companies demand intense scrutiny, and many other subjects, are explored in his essay, "The Adolescence of Technology," which we also discussed in Chapter 11. I also note his work further in the models that follow as "New Models for the Future of Humanity."

- Another novel in the canon is Adam Dorr's *More Than the Sun and Stars,* set in the near future and developed around brain-computer interfaces, a key technology among those we've been exploring. The protagonist, Hoshimi Lancaster, born with severe disabilities, becomes the first subject to receive a full neural implant suite from the fictional company ExoCortex, which links her brain to vast cloud computing resources, enabling her to achieve superhuman intelligence. Initially framed as a medical restoration story—to help her regain normal motor and cognitive function—the book quickly asks, *What is "normal" once you can scale cognition arbitrarily?* As Hoshimi's capabilities grow, she discovers other augmented minds and competing institutional agendas: corporate, military, and activist. It really turns on the heart of *Love Conquers Fear*: Exponential technologies can catalyze "super-abundance," while we need to grapple honestly and deeply with responsibility, dignity, and humanity in a world of radically uneven capabilities.

This was a fun read on brain-computer interfaces and really crystallized for me how powerful they could become. It isn't so much a spiritual read but it is quite moving. I learned about it from the CEO event at MIT where I spoke and referenced in the Introduction. From that point on, I began to refer to brain-computer interfaces as the fourth major force when citing the exponentially accelerating technologies of AI, quantum computing, and robotics, forming the Superfecta.

- From "near" to "nearer" is the clever way sage, investor, and entrepreneur Ray Kurzweil frames *The Singularity Is Nearer*, his 2024 sequel updating his 2005 book, *The Singularity Is Near*. His earlier books, including *The Age of Spiritual Machines*, really put the ball of so much popular discussion of AI into play. The merger of mind and machine, he believes, is nigh and *The Singularity Is Nearer* offers a fresh timeline and more detailed roadmaps. In many ways reflecting the science behind Dorr's novel, Kurzweil doubles down on exponential curves and argues that we're on track for human-level AI in the 2030s, brain-cloud interfaces soon after, and radical life extension as biotech and nanotech converge. The book is a good tutorial on contemporary AI advances (transformers, multimodal models, scaling laws) and maps them onto his long-standing "Law of Accelerating Returns." He suggests exponentially growing generative AI tools like ChatGPT, Claude, and Gemini are a mere glimpse of the fully "spiritual machines" just over the horizon. The more interesting, and controversial, sections are about his core belief that as humans we will gradually offload our memory, reasoning, and even aspects of our selfhood into networks. He believes that the hard boundaries between the biological and the digital will dissolve. I had the chance to meet and speak with Kurzweil after he presented at the April 2024 TED conference, shortly before the release of *The Singularity Is Nearer*. I summarized it in a post shortly thereafter, "The AI-Driven Universe a Blink of the Eye Away." It's a bold and mind-bending book and foundational to understanding the scope and scale of AI and other technologies.

The biggest mindbender from this book for me was Chapter 3 on *Who Am I?* I summarized my wandering thoughts on that in a YouTube video after a long walk with family.

- Computer scientist Dr. Fei-Fei Li's *The Worlds I See: Curiosity, Exploration, and Discovery at the Dawn of AI* is unique: part immigrant struggle and love of family, part sheer grit to succeed on the frontier of academia, and a compelling account of the concrete ways AI is already enhancing our lives and will continue to do so. After her challenging struggle to get an education—including a scholarship to Princeton while helping her family run their dry cleaning business and a doctorate from CalTech—she broke boundaries and glass ceilings in male-dominated industries to pioneer early research in computer vision, including her creation of the massive visual database ImageNet. The book concludes with a reflection on the social responsibilities of AI researchers—fairness, healthcare applications, education—and the dangers of militarization or surveillance. As a result of those concerns, she co-founded and is co-director of Stanford University's Institute for Human-Centered AI. I reviewed the book in early 2024 as, "Attention AI Technologists, It's Time to Merge Intentionality with Philosophy, Ethics, and Law." It's a vital and inspiring book that insists that care, compassion, and moral imagination belong inside the story of AI, not outside of it.

Meanwhile, Li has also gone on to found World Labs, a spatial intelligence company focused on building next-generation AI models that can perceive, generate, reason about, and interact with 3D environments, promising safer buildings, factories, homes, and a more interactive environment for robots. Founded in 2024, her newest venture was valued at $1 billion in its first year, having raised $230 million of funding out of the gates. I encourage you to watch a demo of Marble, World Labs's first product, online.

Journeys, Meaning, and Classic Wisdom

In my recounting of the "guidebooks" I've turned to, I've largely focused on the fundamental nature of reality, the scientific, AI, and the nature of consciousness. But in this last section I want to share the books that explore the terrain of the inner life—the terrain of meaning, purpose, responsibility, and the evolution of the soul. Whereas the first section pushed outward into the cosmos, here the direction is inward, toward the question that arises once we grant that reality is more than material: *How should we live?* These authors, each in their own way, trace journeys that began in crisis or rupture and open into a wider, more spacious understanding of the human experience.

Foremost is JD Messinger's *11 Days in May*, which begins with just such a rupture that JD described in this book's Foreword. We discussed JD and his book in Chapter 10, "Religion." JD's book is as unique as he is—Annapolis grad, nuclear scientist, submarine officer, CEO, television host, writer, and now a good friend. The publishing company he leads, Do More Good® Publishing, is also the publisher of this book. He and I were brought together by Jay Wilkinson, the founder of the Do

More Good® movement, whom I became good friends with after the talk I gave at that same MIT gathering I discussed in the Introduction. Every time I'm called to travel, I look for the synchronicities in why I'm there, spurred to do so by the book *The Celestine Prophecy*, mentioned in Chapter 11, "Risks." For me, that was the relationship I built with Jay; it was clear that he and I were meant to work together.

After reading and being moved by *11 Days in May*, I asked Jay to meet JD and found out, to my surprise and delight, that he was now leading Do More Good® as their Executive Director. On episode 41 of the *Love Conquers Fear* podcast, Jay and I discuss the upcoming Do More Good® "ROI of Why" conference that will help launch this book and the movement to have leaders embrace love conquering fear so that humanity enters the Age of Abundance for All. The conference is inspired by an ethos similar to the Conscious Capitalism movement we discussed in Chapter 9, "Capitalism." Jay's movement is a Nebraska-based initiative focused on helping companies shift from traditional profit-only thinking to a purpose-driven, stakeholder-centric model of business.

JD was on a two-part episode 24 of the *Love Conquers Fear* podcast. In his own book, which he wrote over those eleven days of channeled wisdom, he ponders thirty-six existential questions, including, *Who am I? Why am I here? What is God?* As with Eben's book, JD recorded the audio version in his own voice—both as questioner and responder—so the listener hears the inner dialogue unfolding in real time.

What emerges is a deeply moving encounter between two states of reality that he calls the *World of Form,* with its inherited materialist assumptions, and the *World of Light,* the domain of meaning, intention, and interconnected souls. This deeply moving dialogue between the two dismantles the inherited assumptions of a materialist worldview and replaces them with a cosmology of interconnected souls, purpose-driven lives, and a universe that speaks in symbols and synchronicities. The book's original format mirrors its thesis: that reality itself is a co-created conversation. JD's journey is, ultimately, an invitation for the reader to take their own final exam—to reconsider every major life assumption in the luminescence of *The Light.* In that illumination, we can begin to break the programming imposed by culture, family expectations, false ambitions, and empty pride, and rediscover the deeper intelligence attempting to speak through our lives.

Another profound seeker, Gary Zukav, extends this exploration in *The Seat of the Soul*, a book of many dimensions. Zukav argues that humanity is transitioning from the five-sensory perception we are familiar with to a multi-sensory awareness, a shift in which we begin to recognize ourselves as immortal souls evolving across lifetimes. His framework blends reincarnation, karma, and non-physical guidance into a deeply relational spirituality, where emotions serve as feedback for the question we opened with: *Are we acting from fear or from love?* Choices become tools for soul growth, and relationships become "spiritual partnerships" for mutual evolution. Zukav's tone is gentle, accessible, and universal, making his work a bridge between mystical traditions and modern self-understanding. Where JD offers the cosmic exam, Zukav offers the daily practice, suggesting how to use our power lovingly and how to align our lives with the deeper currents of purpose. I suggest buying this book on both Audible, as Zukav's voice is very soothing and moving, and as a print copy for the many deeply inward exercises at the end. Going through those exercises will be transformational for you and there are many online resources as well.

And then there is Viktor Frankl, whose book discussed earlier, *Man's Search for Meaning*, has long been a personal inspiration for me and one that I discussed at length in my first book, *The Entrepreneur's Essentials*. Unlike most of the volumes I've shared, *Man's Search for Meaning* is not new; it was first published in German in 1946 and in English in 1959. But it is foundational and is essential to bring us back to the ground of human experience with unparalleled clarity and moral force. Frankl's account of surviving four Nazi concentration camps is not presented as a spiritual vision but as a hard-won psychological truth: even in extreme suffering, meaning—in the form of love, future purpose, or the freedom to choose one's inner stance—can determine survival more than physical strength. A psychiatrist, Frankl invented the practice of logotherapy, a means of healing through meaning, that frames the human being not as a creature driven by pleasure or power but by a will to meaning.

Frankl's eternal wisdom and presence alongside JD and Zukav

provides a crucial balance. Even if we accept that consciousness is fundamental, or that souls evolve across lifetimes, the real question remains: *What meaning will I commit to in this lifetime, here and now?* Frankl's insistence on responsibility—on answering the question *To what, or to whom, am I responsible*—connects these timeless inquiries directly to the ethical challenges of our technological and planetary age. Together, these authors form a constellation of wisdom. JD shows how crises can crack open a larger reality. Zukav shows how to live inside that reality with intention and love. And Frankl reminds us that meaning is not just granted from above; it is also created through our free-will choices, especially under pressure. As the universe is alive with consciousness, these works suggest that our task is not merely to understand it, but to participate in it through purpose, compassion, and the courage to choose meaning even when the world grows dark.

Other books in this last third of the canon include:

- *The Alchemist,* Paolo Coelho's fable is a book I've deeply enjoyed, having read some passages many times. If you read it every decade of your life, its deep wisdom will unfold and be understood differently each time as you yourself grow through life's many experiences, whether by suffering or by love. *The Alchemist* follows the shepherd Santiago on his quest for a treasure near the pyramids, guided by dreams, omens, and archetypal helpers (the king, the alchemist, the desert woman he loves). The real "treasure" is the discovery that the universe conspires with us when we pursue our "Personal Legend." The book distills perennial themes—listening to the heart, learning from setbacks, the unity of all things—into a very simple, almost childlike narrative that nonetheless resonates widely. It's less about doctrine than about *felt* participation in a meaningful cosmos. In this broader list of books, it tells us in accessible and straightforward language that outer journeys are mirrors of inner transformation. On Audible, it's a treat as it is read by the voice of the famous British actor Jeremy Irons.

This book helped me through a trying period in my life with the fateful line, "It's always darkest before the dawn." The synchronicities showed up for me over and over again, and I knew I was meant to read this book to get through them.

- Drawn from retreats he led as a psychotherapist and Jesuit priest, Anthony de Mello's *Awareness* is deceptively simple. In this wonderful book he blends Christian language with Zen and Vedanta—which I studied in 2013 just outside of Pune, India, with my wife, Debra—to argue that most of us are "asleep." De Mello, in terms similar to the insights of JD, suggests that we live by conditioning and desires we didn't choose. Awakening isn't about adopting new beliefs. It's about relentless, compassionate noticing—of anger, attachment, fear, and self-image—without resistance. That attention dissolves illusions, leading to freedom and spontaneous compassion. He is unsparing about religion: much of what passes for piety he sees as fear-based bargaining with God. De Mello offers a practical phenomenology of awakening that doesn't depend on exotic experiences or grand cosmologies. He makes the case that most human suffering comes from clinging to beliefs, expectations, and desires, and that genuine awakening requires letting go of these attachments to see reality clearly and live fully in the present moment. I'll be diving into De Mello more in Chapter 13, "The Inner Journey."

- James Redfield's *The Celestine Prophecy*, which I've referenced many times in this book already, is a spiritual classic, a new-age adventure novel that reads like a spiritual take on *Raiders of the Lost Ark*. The protagonist travels to Peru chasing a set of "insights" that, one by one, reframe reality: meaningful coincidences (synchronicity), energy fields between people, control dramas in relationships, and the possibility of collective evolution toward higher vibration. The plot is secondary to the structure: Each chapter more or less *is* an insight plus a scenario to illustrate it. It's been a hugely influential book for people moving beyond traditional religion into experiential spirituality. It's also notable that, even though it was written more than three decades ago, before artificial intelligence was a widely known concept or term, Redfield specifically foresees AI—in the "9th insight"—as a technology of survival and liberation. It's a very optimistic book that I believe foreshadows the Age of Abundance for All with its idea that when enough individuals raise their consciousness, social and planetary transformation will follow. On Audible, there is a bonus interview with Redfield at the end on what he now believes, decades after his book came out. I also recommend Audible for this book because of how much more exciting listening to it makes it—think of listening to it as a really deep movie.

Once I read this book, I started to see

synchronicities everywhere. It was a gamechanger for me and has led to a much fuller and richer life. It will help you see how playful the divine can be. For example, the way *The Alchemist* came into my life on Necker Island, encouraged by Richard Branson and other conscious capitalists, was very on point with *The Celestine Prophecy.*

- *Many Lives, Many Masters* by Dr. Brian Weiss is a pioneering book for the Western audience that empirically makes the case for reincarnation. Weiss, a psychiatrist, describes how a skeptical attempt at past-life regression with a patient ("Catherine") turned into a multi-life therapeutic journey. As she revisits prior incarnations and apparent between-lives states, long-standing phobias and symptoms fade. Weiss also reports channeled messages from "Masters" conveying a spiritual teaching about learning lessons, karma, and the continuity of the soul. The narrative is simple, and it makes reincarnation and regression therapy accessible, mainstream topics. It's an influential bridge between clinical practice and esoteric cosmology that aligns with and mirrors much of what I've learned from Eben Alexander, JD Messinger, and Michael Newton about whether experiential information from altered states can reshape our models of mind and destiny. *Many Lives* is another great book on Audible, as it is both short and read by Weiss, where you can directly hear the awe in his voice.

Thanks to Mark Gober's recommendation, this was the original book that helped me see that reincarnation was real. Then I had friends who told me about the past lives they've experienced both within and outside of psychedelic medicine.

- *Journey of Souls: Case Studies of Life Between Lives*, by Michael Newton, is a critical book as it probes a dimension of reincarnation that I believe is too often neglected—what happens in between. Newton, a hypnotherapist, reports twenty-nine cases of clients regressing not just to past lives but also to the between-life state. This is the spiritual realm where souls review previous incarnations, study, plan, and choose future bodies. The book organizes these accounts into a quasi-ethnography of the afterlife: soul groups, guides, councils, and learning environments. Newton's tone is clinical rather than evangelistic, presenting recurring patterns as evidence for the structure of our lives as a grand cycle of birth, death, and rebirth. It's a major challenge to both materialist science and much of western religion. I found it detailed, internally consistent, and profoundly comforting and clarifying. It gives the afterlife the same descriptive richness we apply to cultures on Earth. *Journey of Souls* is a key checkpoint in your own journey. I recommend the print copy as the voice of the reader is tedious in the Audible recording. I also recommend reading it later in your own soul quest, should you feel moved to do so. This would have been a hard read for my ego to take at the beginning of mine. Stick with the far shorter, *Many Lives, Many Masters*, instead to start.

- In *Signs: The Secret Language of the Universe* psychic medium Laura Lynne Jackson shares a high-level examination of reincarnation, looking into the ways ordinary people experience signs from deceased loved ones and the universe. Most of us, if not all of us, have experienced these signs—I certainly have—repeating numbers, animals, songs, and improbable coincidences. Jackson frames these as intentional communications designed to comfort, guide, and reassure us of ongoing connection. The book alternates between case stories and gentle instruction. It's a guide on how to ask for, notice, and interpret signs without becoming superstitious or obsessive. It's an important book to explore how a sense of participatory meaning can suffuse ordinary life.

In one extraordinary encounter, I was discussing this book with a new good friend, Becca, on an absolutely beautiful Grand Canyon trip in August of 2025, thanks to our mutual friend Arjun Gupta. We had a spiritually curious new friend, let's call him Bobby, sitting in between us. She brought up this book and the improbable signs in the form of pennies that she receives after her father crossed over. Becca told us one improbable story after another. She then asked me about what signs I receive from my parents after they crossed over. I mentioned deer, goats, and birds. At that exact moment it was sunset and a goat appeared on the ridge above. All twenty-seven people in our party looked up at the majesty of it standing against the sunset. It was absolutely stunning. I looked at Bobby and said, "Yeah, the signs are something like that," and then we chuckled with joy. Bobby said, "Ok, I know this sounds weird but I found something while making my camp site." He pulled out a very weathered coin, weathered beyond recognition and mused that maybe it came from a Native American tribe. He and Becca went to the river to clean it off. They came back and . . . guess what it was. Yep, a penny! Bobby's mind was blown and he told this story to the entire group at our checkout. Those signs were meant for Bobby, not just for me and Becca. Sometimes the divine hits you over the head to wake up your heart.

Getting *Signs* on Audible is highly recommended. Jackson's voice will move you to tears of joy in every chapter if it affects you as much as it did me.

- Michael Singer's *The Untethered Soul: The Journey Beyond Yourself* is an important book that describes how to get to know the Self beyond. Singer starts with a simple question: "*Who is the one inside who notices your thoughts?*" He then expands this "inner roommate," his wonderful term for our constant mental chatter, into something larger, as an object in our awareness. Freedom, he suggests, comes from resting our awareness and letting thoughts and emotions pass through without clinging to them. This is a very practical book that walks through clear steps—relaxing around triggers, opening the heart, even facing death—to gradually stop resisting experience. It's remarkably clear. I think of it as a kind of modern *Upanishad* for Western readers. In your stack, Singer provides a psychologically grounded, technique-light approach to awakening that, while philosophical and spiritual, aligns with the contemporary neuroscience of Self. And his voice is beautiful on Audible, I recommend getting it there.

This book gave me the tools to commune with my soul and the divine outside of psychedelic medicine and to let go of anger or fear. It's a very practical and actionable read.

- A compact teaching, derived from his lectures as an audiobook, *How to Surrender to God*, by psychiatrist and former atheist David R. Hawkins explores how to surrender through integrity, spiritual intention, and letting go. Its lecture tracks cover topics like "The Way of Love," "Willingness to Surrender," and "Karma and God." *How to Surrender* is similar to Singer's work but more conventional in its take on the divine. He frames surrender not as passivity but as releasing what he calls "positionalities"—resentment, fear, and the need for control—into an ever-present field of unconditional love. He also emphasizes practical steps: noticing inner resistance, consciously offering it up, and trusting that outcomes handled by God are ultimately more harmonious than those forced by ego. Hawkins' book is devotional and reassuring in its message that we can act responsibly while acknowledging that ultimate control lies elsewhere. While reading this book, I had a profound download about one of my best friend's future that I shared with him. It moved him, and me, to tears.

- A book of a very different frame of all that we've explored is *Zen and the Art of Motorcycle Maintenance: An Inquiry into Values* by Robert

Pirsig. This book's story itself is inspiring. Rejected by 121 publishers, it then sold 50,000 copies in three months after finally being released in 1974, and it went on to sell more than five million copies in twenty-seven languages. It was recommended to me by friends who'd read it years ago, and it's a journey within the journey. A philosophical memoir of sorts, Pirsig braids a motorcycle road trip across America with his son with deep personal reflection. In it, he explores his alter ego's descent into and return from psychosis, all organized around the elusive concept of "Quality." He critiques the subject-object split in Western thought and suggests that pre-conceptual Quality—our direct sense of better/worse, fitting/ unfitting—is primary. The book is less about Zen *per se* than about a lived philosophy where rational analysis and intuitive appreciation are reconciled. The father-son relationship is central to the book's narrative and emotional power. The narrator's relationship with his son leads to a deeply moving self-reckoning and the journey becomes as much about reconnecting as it is about philosophical inquiry. It's also a cautionary tale about what happens when an individual tries to resolve civilizational contradictions in isolation. It's a great book that hit a global chord decades ago and is now being rediscovered. I highly recommend it on Audible as it is superbly read by several voice actors with a lot of great sound effects, like *The Celestine Prophecy* is. It's short but deeply moving. I think of it as the book of my own father, Brian. I read it shortly after encountering his soul in a psychedelic journey.

All the works explored in this chapter sit between two guiding lights, two bookends, if you will. I think of one of them as among the key ancient sources, and the other as the distant horizon of the same great arc of our understanding through these lessons.

On one side stands *The Bhagavad Gita*, a text whose spiritual ancestry reaches so far back that it feels less like literature than—as one sage beautifully put it—a tuning fork for the human soul. Alongside other ancient scriptures to which I've returned throughout my life, including my own Hebrew Bible, the Torah, the *Gita* speaks in a language that defies doctrine. It is a call to presence, to right action, and to the recognition that consciousness—awareness itself—is the true ground of our being. It invites us to remember that the essence animating each of us is not separate from the essence that animates all things. And there are many ways to explore the *Gita*: in translation, as a compendium, or through its long lineage of interpretation.

On the other side of this arc lies *The Lattice*, my own novella I mentioned at the outset of this chapter, a work I wrote as both an imaginative exercise and a synthesis of the lessons and the future toward which they have been guiding me. Where the *Gita* voices the perennial wisdom of our ancestors, *The Lattice* envisions the spiritual and scientific unity of our future. Its protagonist, Dr. Alexander Bliss, discovers what many of the thinkers in this compendium of readings suggest in their own terms: that consciousness is not emergent but fundamental; not an epiphenomenon but a field of relational intelligence woven into the universe. In the world of Dr. Bliss—a future just a breath away from our present—the proof of our interconnectedness becomes the catalyst for a unified humanity and a planetary awakening.

If the *Gita* is an ancient riverhead, *The Lattice* is an imagining of where such waters might ultimately flow. The unity we long for is already here. The future we imagine is seeded in the present and propelled by all the technologies we've discussed. By bridging this ancient wisdom with our Superfecta future, we move from the "hard problem of consciousness" to the practical reality of abundance around the bend.

Taken together these readings illuminate that our awakening is not a destination but a *remembering*—across space, time, and culture—of the truth that has always been ours to claim. I know in my soul that these waters flow toward the Age of Abundance for All.

Communicating with God is the most extraordinary experience imaginable, yet at the same time it's the most natural one of all, because God is present in us at all times ... We are connected as One through our divine link with God.

—Eben Alexander, author of
Proof of Heaven

CHAPTER 13
THE INNER JOURNEY

The greatest tragedy of human existence is not suffering, but unconsciousness. People suffer because they are asleep. Once you wake up, you see that nothing is missing— everything is already given.

—Anthony de Mello, author of *Awareness*

As mentioned in the Introduction, I ended 2024 in Kerala, India, a deeply spiritual place that pioneered Ayurvedic massage, medicine, and holistic health. It is a place to which people travel from around the world—to deprogram, to shed what no longer serves them, and reprogram with what does.

That so many feel the need to travel halfway around the world simply to find quiet tells you something about the world they left behind: conflict, anger, rivalry, jealousy, trauma. There is no doubt that we are in a war today. It is the war for your attention, your body, your support, and your dollars. We've created a version of *The Matrix* by putting profit over people. Deep down, humanity knows it is wrong but we are programmed to believe in YOLO (You Only Live Once). Everywhere we turn, we are programmed to crave toxic food, to support politicians who tell us if we don't it will lead to the apocalypse, to spend on items we don't need (and will bring us but a flash of happiness, nothing lasting), and to be enraged by the latest failures of humanity in the constant barrage of media and social media pushes. If we sit back and become "consumers," just waiting for the latest "push," then these preprogrammed desires and empty promises can quickly take us into despair and depression.

I'm a capitalist, yes. But as we've established in Chapter 9, "Capitalism," I believe in the strain known as Conscious Capitalism. The dated capitalist would argue, "But we give people what they want!" Let's take food as an example. As John Mackey said on episode 30 of the *Love Conquers Fear* podcast, people want foods that are the most addictive and least healthy. Of the 300,000 years that we Homo sapiens have existed, electricity has been part of our lives for barely a fraction—less than 0.1 percent of our entire history. Agriculture only began around 11,000 years ago, which means for around 96 percent of our existence we didn't even have that. And it took a long, long time after that for it to become a globally connected trading system to get fresh blueberries in America almost any time we want.

Not so long ago, we dealt with unimaginable food scarcity that made finding something sweet in nature a truly novel discovery. Now, sugar is abundant. How many foods around the world are just a cultural remix of a cheap starch, like flour or rice, and sugar? Yet we crave sweetness for the few seconds of pleasure on our tongue and then deal with the consequences as that unhealthy fuel gets processed by our bodies' biological engines. But the pharmaceutical complex has come up with a solution to keep the cycle of sugar craving and unhealthy eating going: GLP-1 drugs like Ozempic®, Wegovy®, or Mounjaro®, which around 6 percent of Americans are on today and around 12 percent have tried. Tens of billions of dollars are being made on GLP-1s and that number is projected to triple by 2035. Our biological engines—fueled by addictive sugar and regulated by expensive drugs—are a physical manifestation of the scarcity mindset, which keeps us trapped in a cycle of consumption and repair, ensuring our amygdalae remain hijacked, as we explored in Chapter 1, "Conquering Fear." Meanwhile, the epidemic of heart disease, killing one American every 33–34 seconds, is a stark reminder that this prioritization of profit-over-people has real, lethal consequences. Algorithms are defeating us, killing us one by one. With their massive data inputs and computational power, they know us better than we know ourselves. It's as if we are sleepwalking. Keeping our amygdalae hijacked is big business.

But there is hope. The youngest generation has declining alcohol and social media use. They appear to be more like Neo in *The Matrix*, ready to stop the bullets. They are also deeply connected to and accepting of each other.

My Sacred Path to Waking Up

For years, I was what I now call a successful sleepwalker. As a fifty-year old man in 2022, I had become overly programmed, running a "scarcity operating system" at peak efficiency while winning by every external metric as measured in the World of Form. I was susceptible to being whipped up by political news, excited by pornography, reporting five to six alcoholic cocktails a week to my doctor at annual checkups, and trapped in a comparison machine where the dopamine hit of the next capitalistic award defined my identity. I was playing a game that was too small for the divine spark that we all carry.

Yet I had done a lot of internal work, from studying a key philosophy of living named Vedanta under Swamiji Parthasarathy at The Vedanta Academy in India to being a vegetarian over the past fifteen years. My health biomarkers were great, even with the five to six cocktails per week. I had reformed my workouts and built up to a twenty-four-minute a day plank to strengthen my core and alleviate the common back strains of men my age. I was celebrated as a startup culture leader, and I was leading my sixth company, data.world, that was backed by Goldman Sachs and has consistently been named one of the

Best Places to Work in Austin ever since our inception. I had even been named as one of the Most Exceptional Entrepreneurs by Goldman Sachs that year, at a swanky event with their CEO and top leaders. Given all of this, describing myself as a sleepwalker may surprise many reading this who know me.

One of my closest friends—let's call him Xavier to keep his identity anonymous—had been going to a psychedelic retreat center and telling me about his transformation from 2020 onward. Xavier was the last person I ever expected to have a conversation like this with. He had been very strait-laced growing up, never experimenting with "drugs." I put "drugs" in quotes because he had grown up drinking alcohol at parties. Somehow people always forget to characterize the "dumb molecule" of consciousness modification as a drug, even though more and more studies show that alcohol is nothing more than a poison.

I had taken psilocybin mushrooms and LSD once each when I was much younger growing up in Austin. Both were mostly just weird external experiences for me with strange visuals (I remember the carpet turning into reeds swaying on the seafloor.) But I may have helped save one of my friend's lives during his trip. He was really suffering from his father not being in his life and there were a lot of tears that night. We processed that together in that raw state and he came out of it with some real healing. Looking back, taking psychedelics had been a passing curiosity and it had been over three decades before I considered doing such a "taboo" thing again.

Xavier persisted and claimed to be having conversations with my soul. I thought this was odd at first, but I knew his heart was pure. I had known him for decades and he is one of the most amazing human beings I know. He is a dedicated entrepreneur and the very definition of a conscious capitalist. The conversations he claimed to be having with my soul did indeed sound like some higher version of me. "The heavens have vastly more superior technology than we can imagine on Earth, so far much more superior than AI," he told me my soul had said in one of these conversations. And this was well before the launch of ChatGPT. Yes, I thought it was a bit weird but I started talking to Debra, my wife and soulmate, about it. I finally decided that I should go ahead and try psychedelics again. Why was I afraid of medicines that have been used around the world for thousands of years for transformation?

My first journey was as transformational as they come. As I mentioned in the last chapter, author Michael Pollan's account of his journeys were so profound he used the word *ineffable* to describe the experiences in *How to Change Your Mind* and the excellent four-part Netflix series of the same name. I'll try to describe my experience, but I fear my words won't do it justice as it was an ineffably cosmic experience. I first encountered my mom's soul and felt all of her unconditional love and sacrifice put into me since I was a baby. My mom, Brenda, was the definition of love and being with her soul again moved me to tears of joy. Next, she took me to Debra's soul and showed me that I had indeed married my soulmate. And, as importantly, I was healing something through Debra that I always wanted for my mom. I had encouraged Debra over a decade ago to take trips around the world. I always wanted my mom to have that kind of freedom and not completely dedicate every move in her life to the care of my sister, Brandi, and me. Then, I encountered the soul of my best friend growing up, who appeared to be urgently reaching out to me to contact him. I found this incredibly loving. We had been like brothers after all, though we hadn't talked for over a decade. Finally, I encountered the soul of Brandi and saw all that she had done for my mom as she was going through intensive cancer treatments. There were many other experiences that occurred and it was one of the most beautiful days of my life.

I talked with Debra afterwards and expressed my deep and everlasting love for her as my soulmate. I called Brandi and opened my heart to her and thanked her over and over again for taking such good care of our mom while I was building Bazaarvoice. (At the time, our company had gone public and run up to over a billion-dollar valuation, which was very demanding on me as our CEO.) And I texted my childhood best friend the next day. He replied and we got back together and became close friends once again. I was moved to tears three years later when he told me he was so surprised that God had answered his prayers so quickly. I asked him what he meant by "quickly," and he told me he was in a deep prayer to God the day before I texted him. My body lit up and I got goosebumps. My response after choking back the tears, "Well, brother, you aren't going to believe what I was doing the day before."

I've had several psychedelic journeys since and have had deeply profound connections to God in them. I use the word *God* here because when you are in a psychedelic journey you aren't cognizant of what you are saying and *God* is the word I most often use to describe my perceptions during these journeys. As a Jew, sometimes I use the word *Adonai* or *Hashem* to refer to God in a journey. I've referred to consciousness as "the river." I've referred to Heaven both directly and as "the astral plane." The first time I encountered God in a journey and felt that completely unconditional love and bright radiance, I spent the next day mostly crying tears of joy and deep humility.

One of my journeys profoundly encouraged the launch of the *Love Conquers Fear* podcast and the writing of this book. I include a section of my transcript from that sacred journey here. These are the words that came out of my mouth, directly from my soul. I was not aware of what I was saying and it was a message from the divine through my body. There were no "ums," just a linear progression of these words—no doubt a transmission:

> There's a whole universe out there beyond technology. This universe is so beautiful. Yes, technology is just a part of it. The universe is so vast.
>
> So humbling for us just to be a part of it. God wants us to know.
>
> He wants us to know. It's all so humbling. We are humbled by you, God. So humbled. Please, God, use this body, this life, to achieve what you want. To achieve the best outcome possible for humanity. To achieve the best outcome possible for everyone. To achieve the best outcome possible for everyone I love and care about.
>
> My capacity for love, it knows no

bounds. It's infinite. You can love everyone. You can love everyone.

There is only one God. Only one, God. Only one true God. There is only One. That One is infinite. That One is expansive. That One is the universe. That One is everything. That One is the stars. That One is the cosmos. That One is the galaxies. That One is everything. Everything you can imagine. I am that. That is God. That is God.

Humanity needs to stop dividing. Humanity needs to bring ALL together. There are no males or females, there is just humanity. There is just humanity. Humanity needs to love itself. Stop dividing, stop hating itself, stop hating itself, stop hating itself. All division, all this melted away, it all needs to come together. There is just humanity. There is just humanity. Humanity needs to expand its capacity to love. Expand its capacity to love. That is the most important message, that is a message from God. Expand your capacity to love. That is what you need to do. That is what humanity needs to do. Humanity must love more, humanity must love more.

Maybe I should have given a bigger heart to humanity. Maybe I should have given a bigger heart. But your heart is a mindset. Your heart is a mindset. I gave you big brains. Use your big brains to expand your capacity for love. Use your big brains to expand your capacity for love. If you feel love, if you think love, it will be love.

Come on, humanity, you can do this.

I'm counting on you. God wants you to do this. God wants you to do this, humanity. God is counting on you. Humanity, come on. It has to be in this universe. It has to be this time. Yes, it has happened many times before. But this is it, this is the moment, it's in this lifetime, this lifetime, yes, this lifetime, that humanity will soar, that humanity will finally love all, that humanity will stop the hate, that humanity will stop the division, that the world will be full of abundance. It starts with love. Love is abundance. Abundance is love. Humanity has this capacity. I put it in you. God put it in you.

Life is so precious, so beautiful. Our souls live on forever ... but this body, this lifetime, it matters. We have to achieve higher and higher states. The Buddhists, Hindus, Jews, Christians, they're all part of being right. The higher state ... that higher state is that capacity to love, love for all, regardless of religion, regardless of sex, gender, who cares about gender, it's only humanity, we're all human beings.

That capacity for love ... we have to embrace its infiniteness. It is infinite in every way. Humanity has to embrace that. Infinite love ... that is what God has told us. Infinite love and abundance. There's so much abundance in the universe. We see just a minuscule amount here on Earth ... a minuscule amount. It is all over the cosmos. It's all there for us to enjoy and love.

But it starts with us loving ourselves. It starts with humanity loving itself.

If humanity doesn't love itself, then it's not available to us. But we do love ourselves, God, we will. God, we will achieve that state. We will achieve that state, God. We will, we have to. We achieve that state, we transform humanity to blissful, loving abundance. And humanity will live on throughout the galaxy, throughout the universe. That is the rule of God.

God created us to expand love ... to show that we can overcome fear ... and that we can choose love as humanity. That we all know the answer deep in our hearts. We all know when watching movies, reading books, listening to podcasts. When we're weeping tears, those are tears of love. Yes, we have tears of pain too. Those tears of pain are because we did not choose love. We must choose love all the time. Love really is the answer. We have to expand that infinitely. We have to expand that infinitely. We will. This is God's will. This is what God wants for humanity.

We have to remember behind these divisions, there are very few powerful humans. We all have to grant their power. They're just like us. We have to get out of that mindset, there is no hierarchy. We're all just One, we're all just One. The hierarchy is a connected globe of humans. The ultimate hierarchy is God. There should be no difference among us. We should all just love each other. Ask for abundance. Leave behind our past divisions, leave behind our past fears, leave behind our past scarcity.

Just being One. Just being One, all together, this is what humanity needs. This is God's will.

We're all children of God. We all deserve the best. It doesn't matter what you're born into. We all deserve to have a good life. We all deserve to nourish our soul. This is what matters the most. And this is what matters.

We're all just dust in the wind ... dust going over the sand. Humility is love ... when you have that love you have infinite humility. "The 15th Commitments" is a love language. You practice and you expand your capacity to love.

All are on the same level. Yes, people have their roles, but there is no human higher than the other. It is just humanity ... it is just humanity ... that is God's will. We're all just brothers and sisters.

There is only one God. God is here. God cares about us all ... He loves us all ... and we must love ourselves. We must create infinite capacity to love. That is all that matters. Everything is solved with love ... everything. There are no exceptions.

Oh, the Veda. [Note: this is in reference to the Vedas in India, who created Vedanta.] You are also a child of God. We all are. We all just stand in this, it's essence. We all stand in front of you, God, as but specks of sand. You've chosen to enlighten some of us, but those are the ones that chose to be enlightened by you. You are always available to every human, because we are a part of you. Our souls are but a tiny, tiny, tiny part of you. We are just a tiny fragment of you, God. Just a tiny fragment. We are all just in awe.

Humanity ... humanity needs to cure its arrogance. Humanity is so infantile ... so infantile. My friend was right, there are many beings of higher powers. They chose the path of enlightenment. The technology is abundant to them. It is abundant to us if we choose to love. God has all the technology there is. God has the technology to create universes ... and God only did that out of love. God's infinite capacity to love ... God's infinite capacity to soar ... God's infinite capacity for this universe ... God's infinite capacity for all brothers and sisters. We must all come together.

All just One. All just One. It makes me want to dance. Us all being One. One linear across-the-globe human dance. Joined together. That's the world ... just One. End division. Beautiful. One big chain of humans giving the globe a big hug. One big human hug. One big chain of humans all over the world joined as One. What a beautiful thing that would be. What a beautiful thing. We've seen glimmers of it with one love. Different initiatives, but never all humans together. All means all. All humans together joined in one big love. One big hug. One big forgiveness ... we can do better. This is what humanity needs. That is what humanity needs.

Listen to it. Infinite cosmos. See the

stars like this. Beautiful. I want to see the
stars like this in a desert. Just be able to
look up, laying on the ground, and witness
it all. Just lay there and see them all floating
by. All the stars of the galaxy. That's what
we should do, God. Reconnect with our humanity. Thank you, God. Thank you, God.
Thank you, God.

I know this is a lot to take in, especially if you aren't religious, don't believe in God, or have never had a psychedelic journey. I share it with you as a man who has allowed love to conquer his own fears. It is what I received, and I know I was meant to share it with you and all of humanity.

Outside of these cosmic messages from the divine (as if that weren't enough!) my journeys have been catalysts for deprogramming myself. I stopped watching pornography, realizing it for what it is: a form of scarcity and fear that results in the objectification of others. This habit was part of a broader barrier that had prevented me from seeing the divine spark in all people—one of many barriers women still face despite the remarkable achievements of the past century and a half. We explored those milestones for women, and for all of humanity, thoroughly in Chapter 8, "The Objectification of Women." Seeing my body from above in one of my journeys and hearing my soul telling me to take care of the vessel, and knowing that I had transcended the need for it socially, I stopped drinking alcohol. I was never an alcoholic, but it was time for this bad habit to go. I stopped playing the social media game. I chose to be more curious about politics, not accepting the obvious actions (or inactions) that would be incongruent with the deeply sacred message I received above from the divine, but accepting of others' perspectives. And I stopped letting awards define my identity, appreciating them but not craving them. I also improved my relationships, both with family and friends, and I radiate more love than I ever have before. I notice synchronicities from the divine that my ego would have written off as weird "coincidences" on an almost daily basis. Life is more playful and joyous than before—an experience that evokes the spirit of philosopher James P. Carse, whose 1986 book *Finite and Infinite Games* taught that the highest form of living is not to win the game but to keep it gloriously, infinitely in play. Honestly, I thought I was happy and certainly by society's programmatic definitions I should have been. But if I compare myself today to the earlier me, I'm different. I'm a better version of myself.

Many Ways to Wake Up and Deprogram

I want to stress that psychedelics are not the only path to deprogram and reprogram, and if you go down that sacred path I encourage you to do so both legally and at places that you can very much trust. Get references on the providers. In order for you to be able to let your soul take you to where it needs to, it is really important that you are able to completely surrender to the medicine if you go into a journey.

As I write this I'm at a Siddha Maha meditation retreat. My good friend, entrepreneur, and investor Niraj Mehta, who was on episode 47 of the *Love Conquers Fear* podcast 🦋, invited me to it, and I'm in a place of deep gratitude as a result. I've been doing daily meditations and they have been profound. I've been getting similar messages from the divine in these meditations and we are with a beautiful Siddha Maha master named Guruji.

Meditation is a great way to wake up. The Buddhists have been doing this for thousands of years. In the book *Why Buddhism Is True*, the author Robert Wright quotes Rumi as follows:

> *The thirteenth-century Sufi poet Rumi is said to have written, "Your task is not to seek for love, but merely to seek and find all the barriers within yourself that you have built against it."*

As Wright works his way through the book, he concludes, after exploring how our brains evolved through natural selection, that if all humans followed the path of the Buddha and meditated, they would indeed find that our reality is rooted in love. And, once found, you are on your path to deprogramming and humanity will unite, as I concluded in my science fiction novella, *The Lattice*.

In the book *Awareness* by Anthony de Mello, the entirety is about waking up and deprogramming. De Mello's opening talk, "On Waking Up," sets the premise: Most of us are asleep, living mechanically through labels, attachments, and inherited beliefs we've never examined. His sections on self-observation and attachments are particularly powerful. He argues that what we call love is often dependency, and that real change begins not with effort but with seeing clearly. De Mello is like the comedian George Carlin of spiritual teachers, so I highly recommend listening to *Awareness* on Audible. Comedy can be such an effective truth serum and teacher. My good friend Tero Isokauppila, who was on episode 13 of the *Love Conquers Fear* podcast and is founder and CEO of Four Sigmatic (my favorite protein powder and functional mushroom blend company), recommended *Awareness* to me as his favorite among over 130 books he read when he was on his soul quest.

There is also the path of the stoics, which author Ryan Holiday is a modern day practitioner of. Stoicism and Vedanta are highly aligned and both are great deprogramming and reprogramming tools when practiced regularly.

Religion of course is a well-worn path too. And you can embrace the best of it and reject the dogma that separates us. As Jews, we moved on from sacrificing animals at the altar. Many Hindus have moved on from the concept of the caste system. Many Christians have moved on from the concept of others burning in eternal flames in a place called Hell. And many Muslims reject any form of radical jihadism that treats others as subhuman. The God that I've met loves all, unconditionally. Like any practice shaped by humans, religion has a control mechanism but the common root of love that we explored in Chapter 10, "Religion" and Chapter 12, "What Is Our Reality?" should be embraced as a universal truth. It all leads back to the same beautiful message I received from the divine. If the practice feels wrong in your heart, then trust that it is indeed wrong.

There are many derivatives of religion, such as Helen Shucman's

1976 book, *A Course in Miracles*, which is now available online for free and has become a favorite of a number of friends of mine. Groups gather all over the world to study it and embody its exercises. One of my favorite books to date was recommended by John Mackey. It's named *Love Is Letting Go of Fear* and is by Dr. Gerald G. Jampolsky. It is a succinct derivative of *A Course in Miracles* and it offers many simple exercises.

Whatever methodology of deprogramming you choose, make a commitment to fully integrating it into your life. To help you begin, use this deprogramming checklist as your personal toolkit for the inner journey:

- **Meditation:** Practice mindfulness, observing your thoughts without judgment to anchor yourself in the present moment, rooted in love.
- **Awareness:** Practice "self-observation" by noticing when you are reacting out of old programming or ego rather than conscious, heart-led choice.
- **Stoicism:** Distinguish between what you can control (your own thoughts and actions) and what you cannot (everything else), and focus only on the former.
- **Conscious Religion:** Look past rituals and beliefs that do not serve you to find the core message of the universal root of love in your faith.
- **Active Forgiveness:** When you feel grievance, consciously choose to see the other person's divine spark rather than their mistakes, releasing the past to find peace.
- **Learn Continuously:** Books like the ones I've mentioned throughout this one bring you back to your commitments and eternal truth. I personally revisit them often.
- **Buddy Up:** Surround yourself with friends who will help you on this path, and you will also selflessly serve them.

These practices serve as your defense against being thrown off course by the constant deluge of negative programming coming your way. As I wrote at the outset of this chapter, we are in a war and winning starts with the inner journey by training your mind not to be whipsawed.

You need to develop the strength to protect your body, mind, and money from the constant manipulation machine within which we find ourselves trapped. You'll be all the happier for it. You will become a force for good and love in the world, and you will live in alignment with the divinity of your soul.

Now, imagine if everyone followed that path. It really is possible for all of humanity, in this lifetime, to do so.

One should lift up the Self by the Self; indeed,
the Self alone is the friend of the Self, and the
Self alone is the enemy of the Self.
— The Bhagavad Gita

The most important questions humanity faces are not technical. They are not questions of whether we *can* do something, but whether we have chosen *what* to do—and whether we have the wisdom to measure whether we're doing it.

As I began writing this book, I found myself returning again and again to a simple observation: The world is full of experiments we pay little attention to. While we track the dominant narrative of human progress rushing forward by familiar measures—GDP, market capitalization, quarterly earnings, national power—there are places, institutions, and movements quietly asking different questions entirely. And in some cases, they have been asking them long enough to have answers.

The seven models that follow are not descriptions of utopias. They are unfinished works in progress. They do not agree with one another on everything. But each represents human agency, the act of people choosing to step into the arena. They all represent what I have come to call *New Models for Humanity*—a proof of concept—that a different set of priorities, metrics, and institutions can produce radically different outcomes for human flourishing.

These seven models emerged organically through years of conversation, travel, and reflections with my thought partner and collaborator David Judson, with advisors, with friends, and with the kind of synchronicity that *The Celestine Prophecy* would recognize. They span a Himalayan kingdom that constitutionally prioritizes happiness over growth, a Central American nation that abolished its military and became one of the world's greenest economies, a two-thousand-year-old institution grappling with its role in humanity's future, a new kind of research institute built around the philosophically informed science of consciousness, the accumulated wisdom of Indigenous peoples who have stewarded the Earth across centuries, a technological visionary's blueprint for coordinated global problem-solving, and one of medicine's greatest triumphs—the eradication of two dreaded diseases—as models for what collective humanity can and will achieve. None of them are offering complete answers, but all of them are confronting essential questions.

I highlight Bhutan first because it stands uniquely as a nation that exemplifies *choice*—a whole society that looked at the path the developed world was racing down and said, quietly but firmly: *There must be another way.* In writing about the planetary choices before us, including humanity's unfolding decisions that will set our pace and compass toward the transformation to the Age of Abundance for All, Bhutan's transformational experience is an inflection point that illuminates all the others.

What follows is not a survey of perfect societies. It is an invitation to ask better questions—and to build institutions worthy of the answers.

Let's dive in!

Model 1 – The Education of a King, and the Remaking of a Nation's Measure

You may be familiar with the popular image of Bhutan as "The Happiest Nation on Earth," though that is not a claim this spiritually grounded country makes itself. Resolving the enigma of human flourishing remains a work in progress—even here. Bhutan does, however, claim with great authority that it is asking the right questions about scarcity, want, and fear. And it has put those questions into law.

Bhutan is the only country in the world to constitutionally prioritize well-being and environmental preservation over economic growth. The Bhutanese have built a society founded not on GDP but on a less famous metric: Gross National Happiness, or GNH. And while the seven models that follow explore institutions, movements, and individuals also reimagining what human progress means, Bhutan is singular in that it has done so at the level of an entire nation-state—and sustained it across generations.

My introduction to Bhutan began with a 2016 encounter with Prime Minister Tshering Tobgay at the annual TED Conference, where he was a speaker. The idea of this mystical land lingered in my mind for years afterward. As I began writing this book in 2025, that ember sparked to life—Bhutan crystallized as an essential new model for humanity's future, thanks to my wife Debra.

Pondering Bhutan, synchronicity struck, as that inspiring book *The Celestine Prophecy* would predict, as we discussed in Chapter 12, "What is Our Reality?" One morning Debra walked into my office and said,

"Guess what? I think we should go on a Backroads family trip to Bhutan this December." I smiled. The universe had answered. I then told Debra, an intrepid world traveler, about my plans to write about Bhutan—and why this small kingdom's radical experiment deserves a place in any serious vision of humanity's future. It was a beautiful synchronicity and a fitting end to the last stretch of writing this book.

In conversations with David Judson—my thought partner and collaborator on this book—and other close friends, Bhutan had already crystallized as the first of these seven New Models for Humanity I wanted to explore. I'll get to the labor of love that is building a society on a different set of metrics shortly. But first, the journey itself.

Backroads, for those unfamiliar, is the premier active travel company specializing in small-group biking, hiking, and cultural adventures around the world. As much as we love to explore the world, this wasn't an exotic detour. It was a pilgrimage to meet a society that looked at the path the developed world was racing down and said, quietly but firmly: *There must be another way.*

Flying into Bhutan is quite an experience. As we approached Paro International Airport, one of the world's most difficult airports to land at, we swooped down into a narrow Himalayan valley and could see people lovingly waving at us from their homes on each side. The approach required a precise series of tight, last-minute turns between mountains, and very few pilots are certified to land in such challenging terrain. But our approach felt smooth as we touched down.

Bhutan's place in the world and its unlikely emergence as a compass point for humanity's future is the result of one young man's lived experience, absorbed at a peculiar moment in history, and then translated into policy with careful conviction. In 1972, the year I was born, a sixteen-year-old boy became king of Bhutan under circumstances no one had anticipated. His father, King Jigme Dorji Wangchuck—the third *Druk Gyalpo* and the modernizer who had begun cautiously opening Bhutan to the outside world—died suddenly while on a diplomatic mission in Nairobi. The young crown prince, Jigme Singye Wangchuck, was still a student when he was summoned home. Within days, he was crowned the fourth king of a nation that most of the world could not locate on a map. The transition was neither planned nor gradual. Bhutan had lost the architect of its careful emergence into modernity, and in his place stood a teenage monarch inheriting not just a throne, but an unfinished transformation.

But the young king was not unprepared. His father had curated an education designed to help Bhutan navigate both tradition and transformation. Jigme Singye Wangchuck had attended St. Joseph's College, a prestigious Jesuit institution in Darjeeling, where some of the most fortunate South Asians learn English and the arts of diplomacy. Later, he spent formative years in England at institutions where his studies exposed him to Western governance and economics.

During these years, split between colonial-era schooling in India, 1960s Britain, and his homeland, the young king developed what I would call *outsight*: the ability to see his country not just from within, but as the world would see it, and as it might become.

In the late 1970s and early 1980s, traveling abroad as a young monarch, King Jigme Singye Wangchuck encountered the West at the height of its post-war confidence. He saw cities humming with industry and innovation. And he also saw what lay beneath: the loneliness of suburbs, the alienation of workers, environmental wreckage, and rising mental illness in societies that had, by every measurable standard, "won." He saw people who had "everything" but yet felt empty.

The young king recognized that the global development paradigm, embodied in the Gross Domestic Product (GDP) metric, was an extraordinary achievement in measurement, but a failure in vision. GDP could tell you how much a country produced. It could not tell you whether its people were flourishing, whether its culture was intact, whether its land would sustain future generations, or whether its government was just. So, in the early 1980s, King Jigme Singye Wangchuck posed a question that would reverberate far beyond his mountain kingdom: *What, exactly, is development for?*

Bhutan's answer was a new metric by which to measure a society, Gross National Happiness, that was neither anti-modern nor anti-economic. It was a reframing, an early version of what Silicon Valley would later call OKRs (Objectives and Key Results), a system of intentional goal-setting that measures what truly matters rather than what's merely convenient. John Doerr famously popularized OKRs at Google, mentoring the company's young founders to align their ambitious objectives with measurable key results. Doerr later wrote about this approach in *Measure What Matters: How Google, Bono, and the Gates Foundation Rock the World with OKRs*, which I profiled in my first book, *The Entrepreneur's Essentials.* But where OKRs are optimized for corporate performance, GNH is optimized for human flourishing.

The common DNA between Doerr's OKRs and the king's GNH is simple: both recognize that you cannot pursue what you do not measure, and you should not measure what does not matter. Economic historian and investor Zachary Karabell, a friend and member of the Advisory Board of my last company, data.world, makes the same point in his book *The Leading Indicators*: The numbers we choose—GDP chief among them—don't just describe progress; they end up defining it.

GNH, meanwhile, is formalized around four pillars:

- Sustainable development
- Cultural preservation
- Environmental conservation
- Good governance

These pillars were further broken into nine domains, thirty-three indicators, and 124 variables. I won't detail all of them, but environmental protection was written into the constitution, including the requirement that 60 percent of the country must remain forested, as Prime Minister Tobgay emphasized in his TED talk. Development projects were filtered through GNH impact assessments, rejecting initiatives that would generate GDP growth at the expense of social cohesion or environmental health.

What made Bhutan unusual was not simply articulating these principles, but institutionalizing them. Long before "well-being economics" or "stakeholder capitalism" became fashionable in Davos or Silicon Valley, Bhutan had operationalized a pioneering model that is more instructive than ever for the world as we evolve towards the Age of Abundance for All.

Acting on this distinctly different means of measuring progress, Bhutan—whose entire GDP of $2 billion equals the amount Americans spend on coffee every three weeks—has achieved remarkable outcomes through its evolved civic, and even civilizational, consciousness. As Prime Minister Tobgay explained to those of us gathered at TED, his country has achieved:

- Free education through university
- Universal literacy
- Free and universal health care
- A healthy average lifespan that is nine months longer than that of the average American, according to the WHO

Prime Minister Tobgay spoke eloquently at TED about how Bhutan became the only carbon-negative country in the world—not just carbon neutral, but actively removing three times more carbon than it produces. He shared remarkable facts about Bhutan's deep intentionality in caring for its people and land. His call at the end of the talk to create an "Earth for Life" initiative—becoming deeply intentional stewards for our planet by building on the "Bhutan for Life" model—was powerful and received a standing ovation.

But the most powerful lessons I learned came not from policy documents or constitutional provisions—they came from the lived experience of being there. After visiting several stunning 400-year-old temples and being blessed by a monk in what can only be described as a spiritual experience that moved me to tears, one of our Backroads guides, Kevin, and I were talking on one of the hikes. Kevin has been with Backroads for twenty years and has seen a lot around the world. I asked him what made Bhutan so distinct.

His answer: "Bhutan is a place where you celebrate not just what was but what *is*." We discussed how the temples are used today just like they were 400 years ago. The king and queen set the tone as Buddhists themselves, and it seems like almost everyone in the country meditates and is focused on inner contentment, which especially stood out to me at Tiger's Nest, the most challenging and famous temple to hike to. It's the one you can't miss as you watch Prime Minister Tobgay's TED talk. Going straight up vertically and then finally reaching it after passing a towering waterfall that would have made Lord Elrond from *The Lord of the Rings* proud, I was immediately struck by its incredible beauty. Kevin then told me how every year the Bhutanese still line up past the waterfall to wait to be blessed by the head monk.

As we departed Bhutan with a final blessing from a monk, one of the other travelers said how much she wished she could take the spirit of the Bhutanese people back to America. She was choking back the tears, and I felt it deeply too. What a magical place we had all just experienced. I include Bhutan as a new model for *Love Conquers Fear* for an important reason. The ambitious capitalists among us tell me they worry: When humanity reaches abundance, where will we find our meaning? They joke with me, "Will we all just sit around meditating?" as if that would be the end of the human drive. This is a very American perspective, and remember, the person writing this book has started and grown six companies, including a unicorn, and backed over 150 others, including twelve unicorns, and founded my seventh with the Love Conquers Fear holding company. But they are missing the point. What would be so bad about part of humanity doing what Bhutan is doing today? What would be so bad about focusing on Gross National Happiness, except replacing "National" with "Planetary"? Why not pursue the prime minister's vision shared at TED of an "Earth for Life" initiative?

In this spirit of building systems, not monuments, I note that King Jigme Singye Wangchuck abdicated in 2006 at age fifty-one, after overseeing Bhutan's transition to constitutional democracy. He did not cling to power. The country is now led by his son, King Jigme Khesar Namgyel Wangchuck, the fifth Dragon King, who serves as head of state in a constitutional monarchy. Day-to-day governance falls to an elected prime minister and parliament. Remarkably, that prime minister today is none other than Tshering Tobgay, the man whose 2016 TED talk inspired my original desire to journey to Bhutan in the first place.

I remind my ambitious friends that humanity will never cease to be curious. We will always continue to strive and create. Once Earth reaches the Age of Abundance for All, there is an infinite universe for us to explore. As we unite globally to embrace the inevitability of abundance, we will have a globally coordinated technological takeoff. This is what my science fiction novella, *The Lattice*, is all about. Let's learn what the actual laws of physics are. Can we travel faster than the speed of light? Let's find out! There has never been a reconciliation between quantum physics and classical science. Isn't it time? This isn't just about Earth; this is about the realization of humanity's creativity throughout the universe, powered by love conquering fear.

So, as we measure, let's consider: If one small mountain kingdom could pose different questions—and build institutions around them— what questions are the rest of us still afraid to ask?

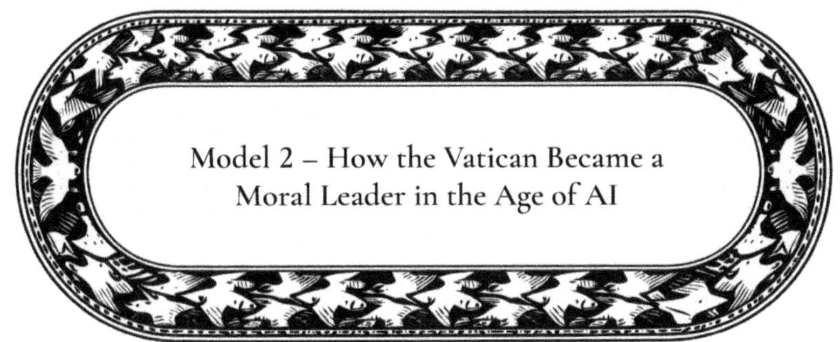

Model 2 – How the Vatican Became a
Moral Leader in the Age of AI

The Vatican perceived early on that artificial intelligence represents an ethical, civilizational, and spiritual challenge, one that transcends the technology itself. In many ways, the Vatican has led and shaped the global moral discussion around AI. I include the Vatican here as a new model because it boldly brings together science and spirituality, which I believe is what humanity as a whole needs to do for love to conquer fear.

The leadership of the Vatican on AI has been remarkable for its breadth. The Vatican's work has extended far beyond the 1.4 billion members of the Catholic Church, or even Christianity more broadly, to include other faith leaders and, notably, business and technology executives.

In a compelling 2023 conversation on his *Tools and Weapons* podcast, Microsoft President Brad Smith suggested to Father Paolo Benanti, the Pope's AI advisor, that the Vatican's work represents a "notion of humanizing technology." Smith continued: "This has manifested itself in many forms, but I think the most tangible has been what is now called the "Rome Call for AI Ethics."

Benanti, an engineer as well as a Franciscan monk, responded: "The power that we have in our hands needs a purpose, needs a direction. And this direction has to be universal because it has to be something that can bring us into the next century at a higher level of quality of human life on Earth."

The 2020 call referenced by Smith was among the earliest Vatican-backed efforts to articulate a moral framework for AI. Promoted by the Pontifical Academy for Life and supported by the late Pope Francis, it committed its initial signatories—drawn from the Vatican, international institutions, and the technology sector—to develop AI that "serves every person and humanity as a whole." While first signed by secular and institutional leaders, the initiative soon broadened its scope and today it has evolved into a platform for interfaith engagement that includes the participation of Jewish and Muslim leaders.

The earliest papal gatherings with tech leaders included meetings in 2016 with Google's Eric Schmidt, Apple's Tim Cook, and Facebook's Mark Zuckerberg. Subsequently, Pope Francis also met privately with Smith, IBM Vice President John Kelly III, Cisco CEO Chuck Robbins, and with Elon Musk, who was joined by four of his sons.

In June 2024, Pope Francis became the first pontiff to address a G7 summit. Speaking specifically on AI, he told the gathering in southern Italy, "We need to ensure and safeguard a space for proper human control over the choices made by artificial intelligence," as he called for banning autonomous lethal weapons.

A month later, Pope Francis sent a message to an AI ethics conference in Hiroshima, Japan. Symbolically, the conference included representatives from all the world's major religions at the Hiroshima Peace Park, ground zero of the 1945 bomb that ushered in the age of nuclear weapons. "No machine should ever choose to take the life of a human being," the Pope told the two-day conference.

In January 2025, Pope Francis sent a message to that year's Davos Forum, where the theme was "Collaboration for the Intelligent Age." In his message, the Pope warned against the "technocratic paradigm," a phrase that would echo days later in the most comprehensive AI guidance issued to date by the Vatican (and possibly more than any major organization to date): "*Antiqua et Nova*: Note on the Relationship Between Artificial Intelligence and Human Intelligence," released on January 28, 2025.

The date was highly symbolic—the annual memorial of St. Thomas Aquinas, a philosopher and towering figure in Church teaching who died in 1274. The timing underscored the theological lineage of the Vatican's argument about human intelligence being fundamentally different from—and irreducible to—machine computation.

While it carries no mechanism of enforcement, as with all ecclesiastical guidance issued by the Vatican, *Antiqua et Nova* stands as a remarkable document of ethical history, reflective insight, and moral suasion. The thirty-page note reviews the history of Church thinking, including the exploration of the very concept of human intelligence made by Aquinas in his thirteenth-century *Summa Theologiae*—the monumental theological work that addressed the nature of reason, the soul, and what it means to think.

Returning repeatedly to the Pope's warning of the "technocratic paradigm" that could diminish humanity, *Antiqua et Nova* emphasizes the critical distinction between human and artificial intelligence. "Even as AI processes and simulates certain expressions of intelligence, it remains fundamentally confined to a logical-mathematical framework, which imposes inherent limitations," the Vatican note reads. "Human intelligence, in contrast, develops organically throughout the person's physical and psychological growth, shaped by a myriad of lived experiences in the flesh. Although advanced AI systems can "learn" through processes such as machine learning, this sort of training is fundamentally different from the developmental growth of human intelligence, which is shaped by embodied experiences, including sensory input, emotional responses, social interactions, and the unique context of each moment."

The profound document, worth reading in its entirety, recognizes the "extraordinary potential" of AI in healthcare, education, and broadly in the economy to improve productivity and democratize access to resources. It also acknowledges the promise of AI "as an aid to human dignity if it helps people understand complex concepts or directs them to sound resources that support their search for the truth."

Still, its language is cautionary against anthropomorphism, the emerging tendencies to vest AI with the characteristics of human love, truth, and empathy. "The vast expanse of the world's knowledge is now accessible in ways that would have filled past generations with awe," the papal note reads. "However, to ensure that advancements in knowledge do not become humanly or spiritually barren, one must go beyond the mere accumulation of data and strive to achieve true wisdom."

Pope Francis died just three months later, on April 21, 2025. His successor, Cardinal Robert Francis Prevost, however, has continued and maintained Francis's rigor and focus on AI. Assuming the papacy on May 8, Prevost took the name of Pope Leo XIII, becoming Leo XIV, precisely because the earlier pontiff centered much of his nineteenth century papacy on coping with the results and ramifications of the Industrial Revolution.

"I chose to take the name Leo XIV. There are different reasons for this, but mainly because Pope Leo XIII in his historic Encyclical *Rerum Novarum* addressed the social question in the context of the first great industrial revolution," the new Pope said in his first formal address to the College of Cardinals just two days later. "In our own day, the Church offers to everyone the treasury of her social teaching in response to another industrial revolution and to developments in the field of artificial intelligence that pose new challenges for the defense of human dignity, justice, and labor."

Since assuming the papacy Pope Leo has sustained engagement with the issues of AI, technology leaders, and policymakers primarily through conferences. His approach has been strategic, using these gatherings to promote the 2020 Rome Call for AI Ethics while asserting moral authority on human dignity and the common good.

The most significant event was the Second Annual Rome Conference on AI, Ethics, and Corporate Governance in June 2025. It brought together representatives from Google, OpenAI, Anthropic, IBM, Meta, and Palantir, along with academics from Harvard and Stanford. AI must account for "the well-being of the human person not only materially," Pope Leo told the gathering, "but also intellectually and spiritually."

At the Builders AI Forum in November 2025, Pope Leo for the first time directly linked AI's promise to the missionary work of the Catholic Church, calling on Catholic technologists and venture capitalists to "build AI systems that help spread the Gospel." The forum, held in Rome, drew some 200 participants from software engineering, venture capital, and Catholic media.

Additional engagements included the World Meeting on Human Fraternity in September 2025, featuring AI roundtables that included MIT physicist Max Tegmark—a sometimes controversial figure whose warnings about existential risks from advanced AI were first sounded in his 2017 book *Life 3.0: Being Human in the Age of Artificial Intelligence*. In December 2025, Pope Leo met with participants in a Vatican conference on "Artificial Intelligence and Care for Our Common Home," where he emphasized helping young people navigate new technologies.

While framed in the language of Catholic theology, the Vatican's work resonates beyond denominational boundaries. Its insistence on human dignity, justice, and love speaks to believers and non-believers alike who recognize that technology must serve humanity—not the other way around. It speaks to the essential message of *Love Conquers Fear: Humanity, AI, and the Age of Abundance for All.*

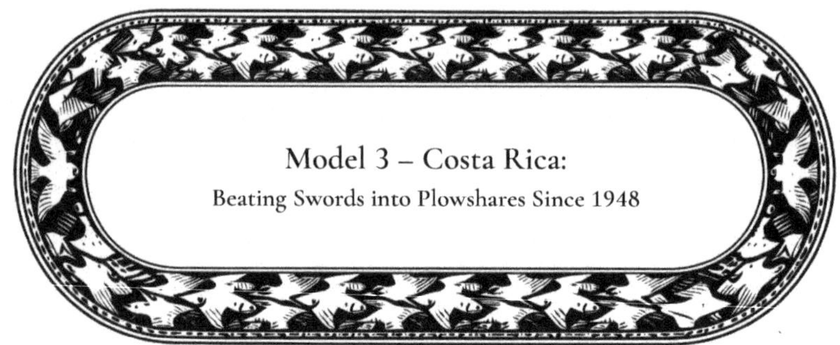

After abolishing its army in 1948, Costa Rica has spent the past seven and a half decades creating universal literacy, a world-class health care system, a nearly fossil-fuel-free power infrastructure, and an economy powered by ecotourism that stands out in a troubled region.

Is it a perfect model easily replicated elsewhere? No. Costa Rica's example results from many exceptional factors. But the Central American nation's distinctive development reflects an ethos that aligns with much explored in *Love Conquers Fear*, particularly the idea of favoring cohesion over coercion, stakeholder principles akin to those of Conscious Capitalism, respect for the natural world, empowerment of women and excluded communities, and a model of shared prosperity.

"Mine is an unarmed people, whose children have never seen a fighter or a tank or a warship," said Costa Rican President Oscar Arias in 1987, accepting the Nobel Peace Prize awarded for his work bringing peace to Central America's war-torn region. "We are an unarmed people, and we are fighting to remain free from hunger."

Importantly, Costa Rica's army abolition, enshrined in the constitution in 1949, did not emerge from abstract pacifism or idealism, but from cold political calculation following a brief civil war. When the losing candidate in the 1948 presidential election attempted to overturn the results, the victor José Figueres Ferrer led an armed uprising that prevailed after six weeks of fighting. While Figueres Ferrer initially governed through a transitional junta with extraordinary powers, he moved the country decisively toward a return to constitutional democracy. But he faced a classic Latin American dilemma: how to prevent the military from becoming a rival power center and coup maker. Costa Rica's armed forces were already weakened and underfunded relative to earlier decades, and Figueres Ferrer moved decisively to eliminate them as a potential political threat, thereby locking in civilian rule.

His choice may have been narrowly pragmatic, but it set in motion powerful positive feedback effects that reinforced the initial decision. Freed from military budget obligations, the government sustained funding for education, health care, and utilities; the absence of an army meant no coups, which strengthened democratic stability and investor confidence; public institutions like the national electricity company were built for universal service rather than strategic control. Over time, national identity became tied to peace, legality, and environmental stewardship, and this created the cultural foundation for Costa Rica's later embrace of ecotourism and conservation.

By eliminating its military, Costa Rica nearly doubled its GDP growth rate, according to a 2025 study, "A Farewell to Arms: The Peace Dividend of Costa Rica's Army Abolition." That complex study, carried out by economists at Texas Tech University and the London School of Economics, estimated the country's growth of 2.38 percent between 1950 and 2010 would have been 1.46 percent had Costa Rica not redeployed military expenditures into development. "We calculate that Costa Rica doubled its per capita GDP approximately every 30 years, whereas, without the abolition of the army and subsequent reforms, it would have taken 49 years to achieve the same growth," wrote authors Alejandro Abarca and Suráyabi Ramírez-Varas. That's an incredible statistic and an act of love conquering fear.

In most of the world, life expectancy closely tracks income. But, like its famous abolition of its army, Costa Rica is a healthcare outlier, wrote noted author, surgeon, and Harvard Medical School Professor Atul Gawande. Visiting Costa Rica as part of research with colleagues for Harvard's Ariadne Labs, Gawande concluded the country's remarkable life span—approaching eighty-one years compared with just under seventy-nine in the United States—is attributable to decades of investment to improve overall public health.

"Even in countries with robust universal health care, public health is usually an add-on; the vast majority of spending goes to treat the ailments of individuals. In Costa Rica, though, public health has been a priority for decades," Gawande wrote in an illuminating 2021 *New Yorker* article, "Costa Ricans Live Longer Than We Do. What's The Secret?" He continued: "Across all age cohorts, the country's increase in health has far outpaced its increase in wealth. Although Costa Rica's per-capita income is a sixth that of the United States—and its per-capita health-care costs are a fraction of ours—life expectancy there is approaching eighty-one years."

As to its energy production, Costa Rica produces more than 95 percent of its electricity from a mix of hydropower, geothermal, solar, and increasingly wind. While Costa Rica is not alone in powering its electrical grid almost entirely with renewable energy—Albania, Iceland, Norway, and Paraguay have achieved similar levels—it ranks among the larger nations to do so with its population of 5.2 million. Albania (3 million) and Iceland (370,000) have much smaller populations, while Norway (5.5 million) and Paraguay (7 million) are roughly comparable in size. But Costa Rica is a standout for its intentionality, asserting control over its hydraulic resources for the general interest as early as 1910. This philosophy hardened into institutional form after 1948 in the same wave of reforms that abolished the army. It began with a creation of a state-run utility with a constitutional mandate for universal electrification, social equity, and long-term sustainability. The infrastructure expanded with rural electrification cooperatives growing in the 1950s and 1960s and from the late 1980s onward, private players entered the market, sometimes controversially, but largely with sustainable models.

The spirit of constitutional reform also carried into early efforts that have evolved into globally renowned biodiversity protection. Again, Costa Rica is an outlier in protecting its environment. While Costa Rica's 20,000 square miles—roughly comparable to the size of West Virginia—is just 0.03 percent of the world landmass, the country shelters some 6 percent of the world's known species. Its ecotourism industry, an estimated 8 percent of GDP, generates over $3 billion annually while protecting 26 percent of the nation's territory as conservation areas. In a *National Geographic* article in 2019, "Four places where humans are living in sync with the natural world," the magazine called Costa Rica's Corcovado National Park "the most biologically intense place on Earth." In a speech the same year at Stanford University, Costa Rican President Alvarado Quesada, called his country a "decarbonization lab." with the aim of eliminating all carbon emissions by 2050. "We are the heirs of a beautiful tradition of innovation and change," Alvarado said, "that's why we're doing this: not because it's fashionable, but because it's an ethical responsibility ... it's not an option, it's a must." That's a beautiful model for the rest of the world to emulate, and it will be more possible to do so than ever as we solve energy sustainably and globally, as we discussed in Chapter 4, "Climate Change."

Other initiatives that grew from the 1948 spirit of reform included extending voting rights to women in 1949, though Costa Rica was not the first in Latin America to do so—Ecuador granted suffrage in 1929. But unlike Ecuador, which restricted voting to literate women until 1978, Costa Rica did not impose literacy requirements. In the years since, Costa Rica has advanced further than most Latin American nations on gender equality, ranking second in the region and sixteenth globally in the 2025 World Economic Forum Gender Gap Index. The country was among the first in the Western Hemisphere to pass gender quota laws for legislative candidates and has achieved cabinet gender parity in recent administrations, placing it among a small league including Canada, Chile, and Mexico. Costa Rica has fully closed its educational gender gap and elected its first female president, Laura Chinchilla, although challenges persist in wage equality and women's workforce participation.

Rights for the 2.4 percent of Costa Rica's Indigenous population present a more complex picture. Barriers to their voting remained into the late twentieth century. But in 1977, Costa Rica declared 24 territories—covering 7 percent of the nation—as "inalienable" and exclusive for the eight Indigenous peoples. While that goal has not yet been fully implemented, it compares with 2.3 percent of the United States designated as Indian tribal land.

Problems remain, of course, and Costa Rica is emphatically not a utopia. Violent crime has surged to crisis levels as the country becomes entangled in Central American drug trafficking. Nicaraguan refugee migration has triggered a xenophobic backlash at odds with the country's egalitarian self-image. High-level corruption has touched nearly every president since 1990. And despite internationally lauded land rights guarantees to Indigenous peoples, two of their leaders defending those rights have been assassinated in recent years

Nonetheless, even with much left undone, Costa Rica's experience offers a new model worth examining as the world enters the Age of Abundance for All powered by the exponential growth of AI, quantum computing, robotics, and BCI—the Superfecta. It is not a blueprint and not a simple roadmap others can follow wholesale. But Costa Rica's distinctive path—choosing education and health care over military spending, conservation over exploitation—demonstrates how foundational choices can compound over decades. That history can help inform how societies balance growth with equity, and technological progress with sustainability, as they build the future.

In short, Costa Rica is a country that my family and I have very much enjoyed visiting over the years. When on the ground and learning first-hand, it was really eye-opening to me that it is powered by such clean energy sustainably, has such a high GDP per person relative to its neighbors, and deliberately chooses to prioritize education and other important development initiatives for its people, animals, and land over supporting a military. It's a model worth studying as we enter the Age of Abundance for All, as in many ways it's an example of love conquering fear.

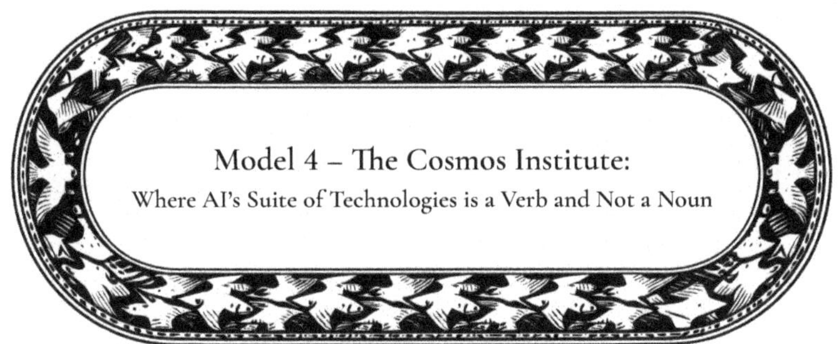

Model 4 – The Cosmos Institute:
Where AI's Suite of Technologies is a Verb and Not a Noun

Every epoch has its sages. Plato and Aristotle guided antiquity's Golden Age in the fifth and fourth centuries BCE. Cicero and Marcus Aurelius shaped thought in the Roman era that followed, from the first century BCE through the second century CE. The Renaissance brought us Erasmus and Descartes in the fifteenth through seventeenth centuries, while the Enlightenment that followed a century later produced the great minds of Kant and Rousseau in the eighteenth century. The philosophy that emerged from the Industrial Revolution of the nineteenth century was defined by John Stuart Mill on one side and Karl Marx on the other. Isaiah Berlin and Jean-Paul Sartre towered intellectually over much of the mid-twentieth century, and the Information Age of the late twentieth century was understood and explained to us by the recently passed Peter Drucker and Alvin Toffler.

We must count among these sages the creators and curators of Eastern philosophy as well. Lao Tzu, Confucius, and the Buddha laid the foundations of Eastern thought in the sixth and fifth centuries BCE. Adi Shankara and the Advaita Vedanta tradition synthesized Hindu philosophy in eighth-century India. The Islamic philosophers Avicenna and Averroes bridged East and West during the medieval Golden Age of the tenth through twelfth centuries. Reformers like Swami Vivekananda brought Eastern wisdom to the modern world in the nineteenth century. Contemporary voices like the Dalai Lama, born in 1935, and the late Thich Nhat Hanh, who passed in 2022, continue to guide millions today.

The canon of Western philosophy, meanwhile, is broadly founded upon the Judeo-Christian tradition. All of these traditions have contributed essential ideas that have shaped civilizations. My good friend Mike Maples—one of the best early-stage technology startup investors in the world—is the author of the brilliant book, *Pattern Breakers*, which I reviewed on Medium as, *To Create the Future of Business, You Need to Break Patterns and See the Future*, and discussed in Chapter 12, "What Is Our Reality?" We then discussed his work on episode 20 of my *Love Conquers Fear* podcast, where he made the case that one cannot fully understand Western civilization without grasping how profoundly that tradition influenced its ethical foundation. This is true regardless of one's personal faith. "For example, Jesus was the person that caused humanity to assume as a given that people had fundamental human rights and that everybody had rights," Mike told me. "Jesus believed in unconditional love. The Western world is undefined without it. And part of what's missing today is people, because they have lost their connection to that, they don't appreciate the importance of those ideas and where they came from." When Mike said this, he was moved to tears. It was a beautiful moment together.

Mike's observation highlights a crucial pattern: Across epochs and across traditions—whether from Athens, Jerusalem, Mecca, or Bodh Gaya—philosophy has not merely interpreted the world. It has supplied the operating systems for civilization.

Which brings us to our own new era: Who will emerge as the defining philosopher of the epoch of AI as we approach the Age of Abundance for All? The candidates are many. Centers and think tanks for AI research now number in the thousands. They span continents: from America to China and Japan; from Germany, to Norway, Slovenia and Great Britain; from Chile to Australia and India. More than twenty institutions have "human-centered" or "human-centric" in their very names, each wrestling with the question of how artificial intelligence should serve humanity.

Yet amid this crowded field, one of the institutions that stands apart for me in its return to philosophy's roots, draws on many thinkers we've discussed throughout *Love Conquers Fear*. This is the Cosmos Institute, which—evocative of Plato's original Academy, named for the Greek hero *Akademos*—calls itself the "Academy for Philosophy Builders." Led by founder Brendan McCord, a serial AI entrepreneur with advanced degrees from Harvard and MIT, the Cosmos Institute bridges ancient wisdom with modern technology.

While the Institute surveys a broad sweep of historical thinkers, three philosophers form its ethical core for the moment of AI: Aristotle, John Stuart Mill, and Alexis de Tocqueville. Each provides essential grounding for building AI with humanist principles. Aristotle taught that human autonomy is essential—that we must remain free agents who think and act independently and not as passive dependents directed by others. Shaped by the Industrial Revolution's upheavals and social cleavages, Mill championed liberty and reason through the free exchange of diverse ideas, advocating for then-radical causes, from abolishing slavery to granting women the vote, as essential to truth-seeking. Tocqueville, a Frenchman inspired by his tour of nineteenth-century America, championed decentralized power and voluntary associations as bulwarks against tyranny from both political centralization and concentrated economic power.

While the Cosmos Institute does not itself use this language, it is useful to understand its work through a particular lens through which AI is seen as a verb rather than a noun. In this sense, Cosmos' uniqueness lies in treating AI and its related technologies not as a "product," but as a suite of capabilities that amplify, accelerate, and catalyze the fundamental capacities for human flourishing.

Just as past eras democratized, industrialized, computerized, and digitized society, we now find ourselves in a moment when society is being "AI-ized." The questions are not "What is AI?" or "How fast can we develop it?" Rather, Cosmos asks: "What does AI amplify?" and "How can it be shaped for our capacities toward human flourishing?"

This is why Aristotle, Mill, and Tocqueville matter. They address precisely what AI acts upon: human agency, intellectual exchange, and the distribution of power. Technology's effects are not predetermined; they depend on what we choose. This core philosophical framing leads

McCord to critique how today's Silicon Valley has lost the moral ambition of such pioneers as Steve Jobs, who saw technology as the means to elevate humanity. As a successful investor and technology insider, McCord's insights are keen and he is direct: Too many of the best minds no longer seek to marry engineering with the humanities, he argues, but rather focus on the data mining and attention-grabbing tools discussed in this book's opening chapter, "Conquering Fear."

McCord identifies three "archetypes" among technologists who dominate a sector that has lost sight of its guiding principles. First is the "puzzle-absorbed," those so obsessed with the next iteration of technology that they can no longer step back and reflect on the broad ends of the technology they are creating. His second set of culprits are the "reductionists," those who would solve all problems with computation, ultimately seeking to make even morality itself "something that is computable." And lastly, McCord criticizes the cohort of "dismissers," those who regard philosophy as nice but antiquated and quaint, a discipline totally irrelevant in today's world. "If AI causes reason to atrophy, humans will slip into dogmatism, intolerance, and persecution," McCord warns.

In McCord's view it's not just an earlier ethos of Silicon Valley that needs to be restored, but the values of past eras when the precepts of philosophy were in fact the North Star for those endeavoring to shape technology to humanity's needs. McCord points to historical examples of this philosophy-technology integration. Benjamin Franklin, familiar to most as one of America's Founding Fathers and the face on the $100 bill, is less well known as an engineer and technologist, McCord points out. Franklin was not merely an inventor of useful devices—from bifocals to the lightning rod—but a thinker who insisted that technical ingenuity be subordinated to civic virtue and human betterment. His experiments in electricity were inseparable from his moral philosophy, as were his commitments to public education, self-governance, and the cultivation of character. Franklin exemplifies a lost model of the technologist, McCord argues: a "moral agent" who understands that progress without reflection risks undermining the very faculties—reason, autonomy, judgment—that make a free society possible. This understanding led Franklin to found the Academy of Philadelphia in 1751, which became the University of Pennsylvania. Education wasn't separate from his technological work but central to it. During my time at The Wharton School, Penn's business school, as an MBA student, I made a habit of stopping at the Franklin bench on campus, drawn to this figure who refused to separate innovation from its human consequences. He remains one of my favorite entrepreneurs precisely because he embodied what McCord argues we've lost—the understanding that the technologist's power demands philosophical grounding.

In eras past, philosophy was the building block of society and civilization at each historical inflection point. It must be again, argues the fast-evolving team at the Cosmos Institute. "Today, we need a philosophy-to-build pipeline," McCord says. "To do this, we're helping create the first AI lab in the world dedicated to translating the principles of human flourishing into open-source software."

The Cosmos Institute is unlike virtually all other AI-focused institutions. It operates as an elegantly integrated ecosystem with three mutually reinforcing components. At its core is the nonprofit academy dedicated to philosophical formation, grounding technologists in enduring ideas about human flourishing, liberty, and civic responsibility. Building on that foundation, the Institute runs seminars, fellowships, and small, rapid grants that translate philosophy into practice by nurturing what McCord calls "philosopher-builders" through the funding of early experiments at the idea and prototype stage. From this educational and research substrate emerges a third pillar: an affiliated venture capital entity named Cosmos Holdings, which provides startup funding to scale the most promising efforts into enduring companies, allowing ideas shaped by philosophy to move, intact, from reflection to creation to real-world impact.

The Cosmos Institute is headquartered in my hometown of Austin and has a collaborative academic presence at the University of Oxford through visiting fellowships, co-hosted seminars, and AI scholarship.

"The next trillion dollars of AI infrastructure will either elevate human potential or introduce a means of perfect control," McCord says of Cosmos' mission. "We are backing the former."

And that's exactly what love conquering fear is all about.

Every age that successfully navigates a civilizational transition does so not by abandoning its inherited wisdom, but by widening the circle of voices it is willing to hear. In this respect, the work of the Cosmos Institute (Model 4) stands out as one of the most serious and admirable attempts to re-anchor AI in philosophy and epistemic inclusion rather than pure optimization. As we discussed, at a moment when AI development risks accelerating faster than our moral vocabulary, Cosmos insists—rightly—that technology without ethical grounding is not progress. It is power untethered.

But there's more. As discussed in Chapter 3, "Geopolitics," the political governance design of the Haudenosaunee Confederacy , better known to many as the six nations of the Iroquois Confederacy, provided much of the philosophical bedrock of America's Constitution and the Declaration of Independence. This illustrates why AI's moral agency cannot be calibrated on the Western Canon alone.

By drawing explicitly on Aristotle, John Stuart Mill, and Alexis de Tocqueville, the Cosmos Institute situates AI within a lineage concerned with human agency, liberty, pluralism, and decentralized power. This is no small achievement. In an era dominated by reductionism and scale-at-all-costs thinking, Cosmos reminds technologists that civilizations are shaped not merely by what they can build, but by what they choose to value.

Cosmos does not claim philosophical completeness. But its very seriousness invites a broader sweep of philosophy than it has yet attempted. It must confront a limitation that matters deeply as we guide AI toward the Age of Abundance for All.

As my good friend Chris Hyams, the former CEO of the jobs site Indeed puts it, tools built entirely on the Western Canon are inherently flawed, biased, and impose at best a kind of cultural tone deafness. "Responsible AI is the single civil rights and human rights issue of my lifetime," says Chris, who now teaches the course "Responsible AI: Language, Technology, Power, and the Future of Humanity" at Austin's Huston-Tillotson University. Huston-Tillotson, or H-T as we call it, is a Historically Black College or University (HBCU) where Chris serves as a trustee and which I have proudly supported as well. Huston-Tillotson is Austin's oldest university, founded in 1875—eight years before the University of Texas at Austin.

The philosophical canon Cosmos draws from—rich, rigorous, and consequential as it is—remains overwhelmingly Western. More specifically, much of it—Aristotle being the exception—is from what historians call the post-Axial Age. This means its scope begins broadly after the period between 800 to 200 BCE when most of humanity's major wisdom traditions emerged, including Greek philosophy, Hebrew prophecy, Zoroastrianism, Buddhism, Confucianism, and Taoism, which all marked a civilizational shift from mythological to ethical and transcendent thinking.

As a result, Cosmos' canon is forged from thought largely developed *after* humanity had already embarked on the path of centralized states, formal hierarchies, property regimes, and coercive institutions. These thinkers wrestled heroically with the consequences of that trajectory, but they did not stand before the fork in the road where other choices were still visible.

If AI is to help humanity transcend fear-based scarcity rather than encode it into digital systems, we must look not only to the philosophers who analyzed modernity, but also to the civilizations that demonstrated viable alternatives long before modernity declared itself inevitable.

We miss many prequels. And one prequel missing from most

of our history textbooks show the influence of the Haudenosaunee Confederacy on Benjamin Franklin, Thomas Jefferson, and the other Founding Fathers through the Indigenous political philosopher, Kandiaronk.

Among the sages Cosmos elevates, Alexis de Tocqueville comes closest to sensing what the modern West had already lost. Traveling through nineteenth-century America, Tocqueville perceived something his contemporaries often missed: that freedom did not reside primarily in laws or constitutions, but in habits of association, local self-governance, and civic culture. He feared that centralized power—whether political or economic—would hollow out these capacities long before tyranny became visible.

Tocqueville, however, arrived at what might be called a hinge point. He could still glimpse remnants of older ways of organizing human life, but he encountered them after European civilization had already declared its path universal. He was diagnosing symptoms, not witnessing the full menu of alternatives.

To find those, we must look further back—and further outward. This prequel is meticulously documented in *The Dawn of Everything: A New History of Humanity,* a book by anthropologist David Graeber and archaeologist David Wengrow which, as one reviewer put it, is "armed to the teeth with fascinating ideas and interpretations that go against mainstream thinking." Crucially, *The Dawn of Everything* explores how the Enlightenment did not emerge in an intellectual vacuum. Many of its most radical ideas—equality, liberty, skepticism of authority, critique of private property—were catalyzed by sustained encounters with the Indigenous societies of the Americas. (Sadly, Graeber died suddenly during the Covid-19 pandemic shortly before the *New York Times* bestseller was published in 2021.)

Kandiaronk, a statesman and philosopher of the Wendat (Huron) people, stands out in particular. Through dialogues with French colonists in the late seventeenth and early eighteenth centuries, including at least one trip to Paris, Kandiaronk articulated major critiques of European society: its obsession with money, its acceptance of poverty amid abundance, its submission to arbitrary authority, and its willingness to trade freedom for security.

These critiques circulated widely in Europe—in particular influencing the seventeenth century judge, scholar, and philosopher known as Montesquieu, best known today for his work developing the concept of the separation of powers—which came straight from Kandiaronk. This Indigenous philosopher influenced Enlightenment thinkers indirectly as well. Yet over time, his origin was obscured, and his arguments were reframed as abstract European philosophy rather than lived Indigenous political reasoning.

This mattered profoundly, because the societies Kandiaronk represented were not theoretical. They were complex, populous, and durable civilizations that had made different choices—consciously and repeatedly.

What Indigenous societies demonstrated, and what modern narratives of progress later denied, is that human beings have always been capable of organizing themselves without permanent hierarchy. Across cultures and continents, Graeber and Wengrow found three freedoms that defined non-hierarchical societies:

- The freedom to move away from authority.
- The freedom to disobey without retribution.
- The freedom to reorganize social relationships.

These were not utopian ideals. They were operational norms. Power was often seasonal, rotating, or deliberately fragmented. Leadership existed, but coercion was constrained. Economic life was embedded in social relations rather than abstract markets.

Most importantly for our discussion, abundance, when achieved, did not automatically produce domination.

This history shatters the assumptions—still embedded in many technological systems—that complexity requires hierarchy, that scale demands control, and that inequality is the natural price of progress. The Haudenosaunee Confederacy provided a living model of political organization that directly influenced Enlightenment thinkers, the American Founders, and most critically Benjamin Franklin. No one absorbed this synthesis more fully than Franklin.

As we discussed on Costa Rica (Model 3), Franklin was not merely a statesman and inventor; he was a translator between worlds. He studied the Haudenosaunee Confederacy closely, admired its federal structure, and saw in it proof that large, pluralistic societies could govern themselves without monarchs or standing armies. The American system of federalism owes as much to Indigenous political realities as it does to Roman law or English precedent.

This lineage matters because it reveals a suppressed truth: Some of the most "modern" democratic ideas were, in fact, ancient—and deliberately forgotten once centralized states reasserted control. This matters because AI will amplify whatever assumptions we embed within it. If AI is trained primarily on Western post-Enlightenment history, it will absorb a worldview which neglects so much: the Eastern philosophies of Buddhist non-attachment, Taoist non-coercion, and Confucian relational ethics that offer essential correctives. Neglecting Indigenous traditions is a further harm, obscuring ways that these societies have structured freedom, meaning, and abundance *without* defaulting to domination.

This is not a call to romanticize the past. It is a call to recover suppressed options. Humanity has faced abundance before. We have chosen differently before. We can do so again.

The question is not whether AI will reshape civilization. The question is whether we will allow the Superfecta to help us remember what we once knew and let that ancient wisdom carry us into the Age of Abundance for All.

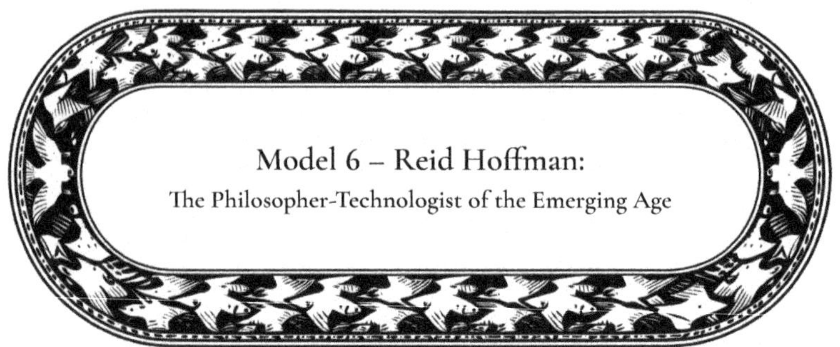

In a famous paper nearly eighty years ago, mathematician Claude Shannon gave modernity a way to think clearly about communication: how to "separate signal from noise." That single act of formalization became the deep substrate of computing, data science, and—by a long chain of technological synchronicities—today's artificial intelligence. Years later, another math genius, investor Edward O. Thorp, remarked that asking about Shannon's impact is "like asking what impact did the invention of the alphabet have on literature."

Today, the closest figure helping us think with similar clarity about AI may be Reid Hoffman—not as the inventor of AI's alphabet, but as an architect of its "grammar," the logic of scale, distribution, institutional adoption, and, increasingly, the ethical and civilizational questions that surround it.

"He is a quintessential Silicon Valley insider and is also known by some as the philosopher of Silicon Valley," said Krista Tippett, author and host of the *On Being* radio program on NPR, in 2023. "Indeed, he studied philosophy at Oxford in his twenties before returning to the U.S. to be part of the digital revolution."

To be sure, Hoffman is in good company. Throughout this book we've discussed other technologists who bring the ethos of philosophy and the humanities into their work. These would include Dr. Fei-Fei Li, co-founder of the Stanford Institute for Human-Centered Artificial Intelligence and the spatial intelligence startup World Labs; Mustafa

Suleyman, who co-founded DeepMind and is now CEO of Microsoft AI; Mira Murati, whose company Thinking Machines Lab is devoted to making AI systems more accessible and human-aligned; Nobel laureate Demis Hassabis, CEO of Google DeepMind; and certainly Dario Amodei, the co-founder of Anthropic AI, whose latest essay "The Adolescence of Technology" I've referenced in earlier chapters. Tristan Harris, the co-founder of the Center for Humane Technology, also makes the list, as we discussed in Chapter 11, "Risks."

While he is not a technologist, the reflection by author Michael Pollan in his newest book, *A World Appears*, which is a deep dive on the nature of consciousness, deserves inclusion in this circle for his work on the alignment of AI with our emerging understanding of consciousness as fundamental. In particular, he masterfully unpacks a great paradox of AI—what takes the most computing power? Abstract and symbolic thought is relatively easy and cheap to compute even though we tend to think of this as at the high end of our intellectual faculties. Sensory motor skills and sensory perception—long regarded as more basic orders of cognition and critical for consciousness—require tremendous computational resources.

Pollan also observes that when he presses the tech community's truest believers hard enough, their arguments tend to collapse into the same three-part catechism. They believe that the future is predetermined and resistance is futile; that digital life will inevitably supersede biological life; and—most revealingly—that this succession should be welcomed rather than mourned. It is, he suggests, a worldview being called in some Silicon Valley circles a "speciesist" outlook: not the familiar bias that privileges humans over other animals, but a stranger and more sweeping conviction that privileges the digital over the biological, the machine over the organism, the artificial over the human. It's less a philosophy than a faith, and it has surprisingly little patience for the human beings it proposes to transcend.

All of these thinkers and doers are pioneers nurturing the AI revolution in responsible ways.

Yet Hoffman's contribution has been distinct. Where others focus primarily on AI's capabilities or safety, he has centered on a different question: how humans and AI systems should work in concert, not just the "alphabet" of AI capability, but the "grammar" of human-AI collaboration, what he calls "orchestration." As a thinker about AI, he is less a prophet than a pattern-reader who is particularly attuned to the ways technologies actually enter the world and reshape human work.

As a technologist, Hoffman's background is unusual if not unique. Born in Palo Alto and raised in Berkeley by attorney parents, he was drawn less to engineering than to ideas, aspiring to become a professor, public intellectual, or writer. As a Stanford undergraduate, he studied symbolic systems—a discipline encompassing philosophy, linguistics, and mathematical logic. His early influences included media theorists like Marshall McLuhan, best known for the maxim, "The medium is the message," and Neil Postman, whose 1985 book *Amusing Ourselves to Death* presciently warned of mass media's corrosive effect on rational discourse—a theme I've explored in the opening chapter, "Conquering Fear," and one which resonates powerfully in today's debates over social media and public thought.

After graduating from Stanford in 1990, Hoffman earned a master's degree in philosophy from Oxford's Wolfson College in 1993, where he'd immersed himself in both analytic and philosophical traditions. Yet even as he engaged with abstract questions of meaning and knowledge, he increasingly doubted whether academic life could offer the direct societal impact he sought.

Upon returning to Northern California in the mid-1990s, Hoffman made an unlikely pivot into technology. He began canvassing Stanford friends for opportunities in the region's nascent tech sector. His entry point was modest: a temporary position in user experience at Apple Computer, which evolved into a role as junior product manager on eWorld, an early attempt at social networking. Though Apple sold the project to AOL in 1996 without achieving sustainable scale, Hoffman had glimpsed the Internet's potential to restructure human connection in profound ways. This insight catalyzed his transformation from observer to builder.

In 1997, barely three years after entering the tech industry, Hoffman founded SocialNet.com. Initially conceived as an online dating service, it expanded into a platform for building all manner of relationships, including professional ones. The venture did not succeed commercially, but it became an intense apprenticeship in entrepreneurship, teaching Hoffman critical lessons about network effects, timing, and product-market fit.

Simultaneously, his Stanford classmate Peter Thiel approached him about joining the board of a fledgling payments company. Hoffman's experience at SocialNet enabled him to help Thiel and his partners execute a crucial strategic pivot—refocusing from person-to-person payments to processing transactions from customers to merchants. By 2000, Hoffman had folded SocialNet, returned his investors' capital in full, and joined what was now PayPal as Executive Vice President and Chief Operating Officer. When eBay acquired PayPal for $1.5 billion in 2002, Hoffman's share provided the capital to pursue his most influential creation.

His experiences at both SocialNet and PayPal had crystallized a fundamental insight: Professional identity would become central to the online sphere. In December 2002, he founded LinkedIn in his living room. When the site launched in May 2003 with 4,500 members, few if any predicted it would fundamentally reshape how professionals build careers and companies find talent. Today, LinkedIn operates with over *one billion members* across all but a handful of the world's 197 countries, territories, and disputed or semi-sovereign entities like the Vatican or Kosovo.

But Hoffman's influence extended far beyond LinkedIn. His share of the PayPal proceeds funded a prescient portfolio of angel investments that helped define the social web, including Facebook, Airbnb, Groupon, Zynga, and Flickr. When approached to lead Facebook's first funding round as he had done for the early social startup Friendster, Hoffman was impressed with the plan but felt that taking a prominent position would conflict with his leadership at LinkedIn. Instead, he introduced founder Mark Zuckerberg to Peter Thiel, who provided the initial angel investment of $500,000. Hoffman has called this "the most expensive decision" of his business career.

In 2009, Hoffman joined Greylock Partners, where he became

a partner and helped shape the firm's early stage investing strategy across consumer internet, enterprise software, social networks, and AI-driven technologies. His leadership of Greylock's Series A investment in Figma—then an unproven, browser-based design platform—reflected a personal conviction grounded in his long-standing theory of networks and collaboration. Where others saw a niche tool, Hoffman recognized infrastructure for collective intelligence: software designed to coordinate creative work across teams and contexts. In this sense, Figma was an orchestration platform before Hoffman began using the term. Adobe attempted to acquire Figma for $20 billion in 2022, but regulatory concerns terminated the deal. When Figma ultimately went public in 2025, its market value vindicated Hoffman's early judgment, demonstrating that enduring advantage comes not from isolated tools, but from systems that choreograph the work of many actors—first humans, and increasingly humans and machines—into shared agency at scale.

By 2016, when Microsoft acquired LinkedIn for $26.2 billion (the largest acquisition in Microsoft's history at that time) Hoffman had become something more than a successful entrepreneur and investor. He had emerged as Silicon Valley's most articulate theorist. His 2018 book *Blitzscaling*, co-authored with Chris Yeh, codified the growth strategies behind companies like Amazon, Facebook, and Google, offering entrepreneurs a roadmap for achieving rapid, massive scale in competitive markets.

In 2023, Hoffman co-authored *Impromptu: Amplifying Our Humanity Through AI*, written with GPT-4 itself, a meta-exploration of AI's potential to elevate humanity across education, business, and creativity. In 2025, he authored *Superagency—What Could Possibly Go Right with Our AI Future,* which I reviewed a month later in "How AI Will Help Us Find the Signal Amid the Noise of the Exponential Age." I've explored his ideas on AI risk, which were formulated in that book and his so-called "Bloomer" approach to human-centric mitigation in Chapter 11, "Risks." These books exemplify Hoffman's unique approach of not merely observing technological change, but actively engaging with it to understand its implications. The philosophical training that once seemed tangential to his tech career had, in fact, become its foundation.

Where others saw only markets and code, Hoffman perceived the underlying patterns: how networks amplify human potential, how platforms restructure social reality, how ideas propagate through digital systems. His synthesis of philosophy and entrepreneurship positioned him to address the defining questions of the AI age with unusual clarity.

In a May 2024 speech titled "Humanity's Hegelian Golden Braid," delivered upon receiving an honorary doctorate from the University of Perugia, Hoffman articulated his concept of *Homo techne*—a reframing of humanity's very identity. It's a concept I've since quoted many times, including in a reflection on democracy and global challenges inspired by my twenty-fifth class reunion at the Wharton School. Rather than defining ourselves primarily as *Homo sapiens*, thinking beings, Hoffman argues we are fundamentally toolmakers and tool users who evolve with and through our creations. "Through the tools we create, we become neither less human nor superhuman, nor post-human," he told the audience. "We become more human."

This insight crystallizes Hoffman's contribution to the AI discourse. Technology is not something separate from humanity that threatens to diminish or replace us; to the contrary, it is the mechanism through which we have always realized our fuller selves. From language to the printing press to the internet, each technological advance has amplified human capacity and reshaped civilization. AI represents not a departure from this pattern but its continuation.

If AI is indeed another amplification tool rather than a replacement threat, this reframes the central challenge. This is why Hoffman's concept of "orchestration" matters. Just as Shannon provided the mathematical tools to understand information transmission, Hoffman is articulating frameworks for understanding AI's integration into human organizations. As Hoffman sees it, the question is no longer simply what AI can do, but how humans and AI systems should work together, addressing critical issues: Who decides what, when human judgment supersedes algorithmic recommendations, and how do we preserve human agency while leveraging machine capability?

Seen this way, Reid Hoffman's importance lies less in any single forecast about AI than in the conceptual perspective he brings to its development. Across his recent work—and most concretely with his creation of Reid AI—Hoffman treats artificial intelligence as a problem of coordination and augmentation rather than replacement. Reid AI is a digital doppelganger of his perspective, designed to engage users in dialogue based on Hoffman's accumulated body of writings, interviews, and talks. It's a living experiment in orchestration: Intelligence emerges not from machines alone nor from humans alone, but from systems that align human judgment with machine capability. The central task of the AI era, in this framing, is not maximizing computational power but designing cognitive and organizational architectures that amplify intelligence while remaining intelligible, accountable, and responsive to human direction.

This is a core premise and goal of *Love Conquers Fear: Humanity, AI, and the Age of Abundance for All.*

As we approach the Age of Abundance for All, our New Models for Humanity will be many, some of which we cannot yet even imagine. For others, the creation and innovation will come with a glance at the playbook of what scientists call biomimicry: the design of systems inspired by the natural world. Imagine the Wright brothers observing birds before that famous day at Kitty Hawk. Recall the Swiss engineer George de Mestral who studied the tiny hooks of burrs lodged in his dog's fur to invent Velcro. Celebrate the Japanese engineers inspired by the kingfisher's streamlined beak to redesign their bullet train's nose—silencing the thunderous sonic boom that had rattled residents up to 400 meters away whenever trains exited tunnels at 200 miles per hour.

As an exercise in conceptual modeling, consider how we typically imagine the way societies plan, build, or organize. Our minds tend to picture what is visible: trees and forests, mills and transport systems, and other man-made institutions and structures rising above the ground. What we neglect, too often, are the root systems—the vast, unseen networks that distribute nourishment, transmit signals, adapt to stress, sustain life, and nurture rebirth when surface conditions turn hostile.

One of the clearest societal examples of such an organizational root system at work—one now largely forgotten—lies in the global efforts that, half a century ago, eradicated smallpox and have since nearly eliminated polio. Before this planetary collaboration united American and Soviet scientists with tens of thousands of local field workers worldwide at the height of the Cold War, smallpox alone killed an estimated two million people each year, while polio paralyzed hundreds of thousands, many for life. Yet even as enmity, an arms race, and profound ideological division defined the geopolitical surface, a quiet, distributed, trust-based network of scientists, health workers, logisticians, and diplomats repeatedly succeeded where formal institutions and political alignment might have been expected to fail. The eradication of smallpox and near-elimination of polio reveal a hidden truth: Humanity's greatest breakthroughs emerge not from centralized power, but from decentralized networks of trust operating beneath political conflict.

Dr. Peter J. Hotez, Dean of the National School of Tropical Medicine at Houston's Baylor College of Medicine, has documented how these campaigns built upon each other through trust established beneath the surface of Cold War hostilities. Between 1956 and 1960, at the height of tensions—just after Sputnik and the first Soviet hydrogen bomb test—American virologist Albert Sabin and Soviet scientist Mikhail Chumakov received diplomatic permission to jointly develop and test the oral polio vaccine on millions of Soviet children. This successful collaboration laid the groundwork that proved invaluable when Soviet Deputy Health Minister Viktor Zhdanov proposed global smallpox eradication to the World Health Organization's governing body in 1958, promising substantial Soviet support. From 1962 to 1966, the Soviet Union pioneered freeze-drying techniques and provided 450 million doses of smallpox vaccine to support the campaign led by America's Dr. D.A. Henderson at the WHO, while the United States supplied crucial financial backing. The trust earned through polio cooperation was redeemed in the fight against smallpox.

That tradition of innovation continues today. The battle against polio in remote regions has been aided by inventions like the Arktek Passive Vaccine Storage Device, developed at Intellectual Ventures Lab by inventors, including my friend and colleague Pablos Holman, mentioned in Chapter 3, "Geopolitics." Using spacecraft-grade insulation, the Arktek keeps vaccines cold for over a month without electricity, thus solving the "last mile" problem in places like Nigeria where the cold chain breaks down and enabling vaccination of communities once beyond reach of the biomedical supply chain.

"The eradication of smallpox and the control of polio were made possible because scientists from opposing nations were willing to work together despite profound political and ideological conflict," Dr. Hotez wrote in an essay for the *Journal of the American Medical Association, JAMA.* "Global politics have clearly shifted since the time of Sabin and Chumakov, and the multipolar interconnected world of today is far more complex than ever before. However, the principles of vaccine diplomacy remain intact."

Shortly before I began this book, Dr. Hotez spoke at the Burt Kunik JAMen Forum at Austin's Jewish Community Center, where I serve on the advisory board. Dr. Hotez's global work on neglected diseases continues to save countless lives, and his reference to this model of beneath-the-field-of-vision service to humanity is one that deserves to be remembered.

I cite this example and the biomimicry insights yielded by root systems—about which we are learning so much, including their means of communication, nutrient transport, and even water sharing among species at times of drought—as we prepare for the advent of the Superfecta of artificial intelligence, robotics, quantum computing, brain-computer interfaces and other technologies amid all the planetary challenges we've discussed. As we've seen in much of the debate around AI, the temptation is once again to search for solutions in visible architectures of power: treaties, summits, regulatory regimes, and centralized authorities. To be sure, these have their place, but the lessons of smallpox and polio diplomacy suggest that enduring progress often depends just as much on what lies beneath: networks of shared technical standards, trust among experts, interoperable data, professional norms, and heart-driven collaboration that operates below the level of dogma, ideology, and national borders. With challenges that no single institution can govern and no nation can solve alone, resilience may hinge less on who commands the forest canopy than on whether the root systems beneath it are healthy, connected, and allowed to grow.

The smallpox vaccine is the oldest inoculation we have. British physician Edward Jenner 🦅 proved its efficacy in 1798, yet uptake remained slow amid public suspicion and side effects that were painful even as the vaccine saved lives. By the twentieth century, embrace of the vaccine grew in Europe and in America, where it was used almost exclusively by militaries to protect battlefield health. Initially, injection methods were rudimentary and painful, and side effects were substantial. But by the early twentieth century, mass vaccinations for civilians grew in piecemeal fashion after the United Kingdom was the first to mandate child vaccination beginning in the 1850s.

Polio vaccines came much later, beginning with Jonas Salk's 🦅 injected vaccine, proven effective in U.S. trials in 1955. But, as mentioned above, it was Sabin's oral vaccine—tested in the Soviet Union between 1956 and 1960 through the landmark collaboration with Chumakov—that proved cheaper and easier to administer in poor countries and became the foundation for global eradication efforts.

Although debate about vaccines obviously resonates today, as legitimate concerns often compete with fear-driven misinformation, this tale of two disease eradications has stood the test of time. For most people living in today's world, both smallpox and polio remain at most a dim memory.

SMALLPOX ERADICATION
Reference timeline

1958 Viktor Zhdanov proposes a ten-year global eradication campaign to WHO; USSR commits 25 million vaccine doses.
1962–66 USSR pioneers freeze-dried vaccine and supplies 450 million doses; United States provides support.
1965 U.S. pledges major assistance; WHO establishes Smallpox Eradication Unit.
Jan. 1967 Programme launches (Dr. D. A. Henderson): 10–15 million cases; 1.5–2 million deaths/year; budget $2.4M (later ~$23M).
Key innovations: Freeze-dried vaccine; bifurcated needle.
1969 17 of 21 West/Central African countries smallpox-free.
1971 Western Hemisphere free of endemic smallpox.
1973 Endemic in five countries only.
26 Oct. 1977 Last endemic case (Somalia).
May 1980 WHO declares eradication.

POLIO NEAR-ELIMINATION
Reference timeline

1956–60 Albert Sabin collaborates with Mikhail Chumakov during the Cold War to develop the oral polio vaccine.
Tested on 10 million Soviet children; later administered to 100 million people under age 20 with diplomatic approval.
Early 1960s Safe, effective oral vaccine enables global mass-vaccination campaigns.
1988 Global Polio Eradication Initiative launched by WHO, UNICEF, and Rotary International; 350,000 cases in 125 countries.
2000s Polio eliminated from all but three countries: Afghanistan, Nigeria, Pakistan.
Today Fewer than 100 cases annually worldwide; polio on the brink of becoming the second disease eradicated.

Today, as living systems teach us so much about how biomimicry might help us reimagine our fast-transforming world toward the Age of Abundance for All, the subterranean root logic takes on even more cogency and urgency in materials science, in quantum computing collaboration, in biotechnology development, and more.

A decade ago, at the annual TED Conference, I had the chance to listen to Canadian ecologist Suzanne Simard. In her talk, "How Trees Talk to Each Other," she pointed out that after thirty years of research in forests, she had discovered that through so-called "mycorrhizal networks"—fungal threads that link root to root—trees share carbon, nutrients, water, and even defense signals. Mother trees nurture seedlings through these hidden highways. Trees recognize their kin and support them preferentially. What appears above ground as individual organisms competing for light turns out to be, beneath the surface, a system of profound cooperation.

"Underground there is this other world, a world of infinite biological pathways that connect trees and allow them to communicate and allow the forest to behave as though it's a single organism," Simard told us. "Forests aren't just a bunch of trees competing with each other, they're supercooperators."

Simard's forests and the planetary collaboration that eliminated two deadly planetary scourges share a fundamental insight: Resilience emerges not from isolated strength but from hidden networks of mutual support, or supercooperators, to use Simard's phrase.

All of humanity must become a supercooperator as we approach the Age of Abundance for All. Love must conquer fear.

As the opening sentence of the purpose statement for the 1968 *Whole Earth Catalog,* Stewart Brand wrote, "We are as gods and might as well get good at it." Steve Jobs was a huge fan of Brand, and I certainly am too. Brand's belief was that humanity already had vast power to shape our Earth through technology and culture and that we should learn to use that power wisely and responsibly. This was in 1968, four years before I was born. It was a year before the first internet message was sent via ARPANET on October 29, 1969 and well before the invention of the World Wide Web, mobile devices of supercomputing power, and ubiquitous social media, and *decades* before the emergence of popular AIs like ChatGPT, Claude, and Gemini. Brand once famously said, "Once a new technology rolls over you, if you're not part of the steamroller, you're part of the road." That is the fear that my friend and technology ethicist Tristan Harris was trying so passionately to express at TED in 2025, which we covered in Chapter 11, "Risks." It's that dystopian feeling that is grafted into our collective psyche—the binary belief that we are either the victims of the machine or the drivers of it, which we also covered extensively in Chapter 11.

But there is a third option. We don't have to be either the steamroller or the road. By stepping out of the scarcity mindset and embracing the lattice I described in my science fiction novella of that name, we become the architects. We realize that the "steamroller" is only a by-product of our own fear, and that we have the power to design a system where technology serves as a foundation for human flourishing rather than a force that flattens it.

It is certainly the case that humanity stands at a fork in the road. I am 95 percent certain that we will make it to the Age of Abundance for All. As I outlined in Chapter 2, "Abundance," the country of Chad could become a place that no one living there today would be able to imagine. Whether that transformation takes place depends on us, and by us I really do mean all of us. As my good friend, entrepreneur, and author Byron Reese said on episode 12 of the *Love Conquers Fear* podcast, "If all of us practiced kindness every day we would all be just fine." Byron is 100 percent certain that we'll make it, barring some asteroid hitting Earth like you see on those infamous bumper stickers that roll around during election season that read something like, "Giant Meteor 2028—Just End It Already!" When you see those and laugh it makes you uncomfortable deep down. It is that gnawing feeling that some of humanity has given up on ... us. On ... each other. On ... ourselves.

As I experienced and discussed in Chapter 13, "The Inner Journey," God wants us to make it and loves us all, unconditionally. We all have a part of God within us, and our souls are quantumly woven into the very fabric of consciousness. So why can't we be like God and love each other in the same way that God loves us? When you put two babies next to each other in a crib, they aren't thinking to themselves *I'm a Muslim and you're a Jew; I'm a man and you are a woman; I'm from the Brahmin caste, you're from the Shudra caste; I'm a Republican and you are a Democrat* or any of the other nonsensical programming that is telling us what is and is not "permissible," which we quickly inherit to categorize and divide us, and telling us how we should think about the 0.1 percent that separates us rather than thinking about our 99.9 percent identical DNA shared by all the people around this globe in our rich cultural tapestry.

As I've explored this topic so deeply on my soul quest, I really believe that it all boils down to this one message: *The heart must conquer the head.* As neurosurgeon James Doty learned in his book *Into the Magic Shop*, when we lead from the heart it is truly hard to go wrong. Doty's big fall came when he forgot to do so, and then he learned Ruth's ultimate lesson (Ruth was the woman that came to his rescue when he was twelve-years old and suffering mightily.) I could hear Doty's tears on Audible while he read this. It led him to start a center named CCARE (Center for Compassion and Altruism Research and Education) at Stanford University to advance scientific understanding of compassion, altruism, and kindness. His Holiness the Fourteenth Dalai Lama, Tenzin Gyatso, is recognized as the founding benefactor.

When Debra and I studied Vedanta under Swamiji Parthasarathy, starting in 2013, we were educated on how the mind is like a child when ungoverned, a nod to his formative book *Governing Business and Relationships.* The mind constantly worries about the past and has anxieties about the future, missing the only moment that matters right here and now—the present. And here we are in the World of Form being whipsawed by an onslaught of programs designed and refined to tempt the mind. While the heart knows, the mind falters. The soul is found in the heart, not the head. When you express deep gratitude, where do you place your hand?

Our ability to make the Age of Abundance for All a reality comes down to the U.S. and China partnering. It's such a hard thing for most to imagine given the difference in our systems, but as the two most sophisticated world powers, those countries—and ultimately the human beings in them letting those in control be in control—will determine our collective fate. This race to create AGI is based on hope but it is also dependent on a zero-sum mindset. The corrosive thought is that if China gets AGI before us, then it may just be all over for the U.S., and vice versa. But China has been offering what I consider an olive branch with DeepSeek and its open-sourced nature, just nipping at the heels of the world's most powerful AI models.

The eastern mindset is more collective, while the western mindset is more individual. It is the blend of both that will form the basis for humanity's collective success. We are all one—both spiritually and planetarily—and we need to learn from each other as opposed to being programmed to fear each other. If you approach someone with fear, how much are you going to learn from them?

I don't have all of the answers, no one does, but I do know this: Byron Reese is right... if we approach each other as Pope Leo emphasized at the *World Meeting on Human Fraternity* in September 2025. When the Pope said that we should see the face of God in everyone, including the poor, the lonely, and even enemies, as part of building a "human alliance" of care and trust, his words really moved my heart. And to think that 2025 started out with the most foundational report on AI that I've read to date ... from the Vatican, as we discussed in Model 2 of "New Models for Humanity." Here was spirituality and science coming together, led by the clergy. Imagine what thoughts were on Galileo's mind as he stood before the Roman Inquisition having committed a single offense. His crime of heresy was that he had looked through a telescope and reported what he saw.

Our future should be one of abundance that we can't imagine today. Around 250 years ago, only about 10 percent of the global population knew how to read and around 80 percent of the global population lived in extreme or subsistence poverty. Time is being compressed now with this exponential-of-exponential curve that we find ourselves in. The exponentially accelerating forces of the Superfecta—AI, quantum computing, robotics, and brain-computer interfaces—are a *forcing function* for humanity's fateful choice. Are we bold enough to choose Abundance for All? Change is hard and the amygdala is strong. The programming to hack our bodies, minds, and wallets is all around us, and sadly profits *are* being put over people. Can we be like Neo in *The Matrix* by finally seeing through the code of fear that has governed us for so long?

Yes, we can. I believe in us. *I love us.* We can be the infinite love for each other that God is for all of us. We are all part of the One. We are all part of God. Consciousness *is* fundamental. The actual reality beyond this Earth School is more beautiful than we can remember when under assault. Our souls know that place of Light and Love, and it is up to us to seize our future together, to usher in the age where Heaven and Earth merge, where humanity ascends to all new heights. Heaven is cheering us on. The divine wants us, and I mean all of us, to win together.

Join our movement.  Love *must* conquer fear. We must do it together, as one big human family, as Team Humanity. Humanity needs an upgrade, and I have no doubt that we are in an evolutionary moment right now. The consciousness awakening is underway. Flow with it, don't fight it. Soar! Everything is going to be more magical than we can imagine or that our ancestors, those *Homo sapiens* of 300,000 years ago on this 4.54 billion year old Earth, could fathom.

There are many great thinkers stepping up to the plate, such as Demis Hassabis, Mustafa Suleyman, Dr. Fei-Fei Li, Mira Murati, Reid Hoffman, and Eric Schmidt. But being a "giant" in this moment isn't reserved for those building the models; every one of us is being called to act as a vital node in the Lattice.

To be a giant in this age is to choose love over fear in every decision, every line of code, and every interaction. Our better angels must win, and that victory starts with our own commitment to Team Humanity. This is when history is made. It's this lifetime, for all of us.

Onward to the Age of Abundance *for All*!

With deep gratitude and infinite love for all of humanity,
—Brett Alexander Hurt

Afterword

In Brett Hurt's ambitious work, *Love Conquers Fear*, he not only covers the gamut of societal issues, but he puts a metaphysical "cherry on top" by examining the nature of reality. That is, he discusses the overarching container for these worldly matters.

Actually, calling it a "cherry on top" undercuts its importance. This metaphysical foundation can be more accurately described as "vital." Why? Because in order to have a perspective on why geopolitics, healthcare, energy, education, or *anything* matters, there needs to be a perspective on who we are and why we're here in the first place. Why do we even exist? What is this world? Brett expresses his view early in this book: "I want to be clear upfront on what I believe: consciousness is fundamental and is the very nature of reality itself, with love encoded as its highest value."

I'd like to double-click on this outlook and also lend it some credibility. Without this underpinning, the notion that "love conquers fear" might sound like an idealistic, utopian concept. Quite to the contrary, Brett seems to be leading us toward a fundamental truth about our reality, whether we like it or not. To some, it might sound too comforting to be true. But just because something sounds comforting doesn't mean it's not true.

Conversely, the current mainstream view of reality is not comforting. And according to many scientists, it is true. They claim that there was an event that started the universe, typically theorized to have been a "Big Bang," which occurred roughly 13.8 billion years ago. Then, units of matter (atoms) were buzzing around space, and they slammed into each other like a series of billiard ball collisions, which eventually led to the evolution of human beings who existed on a flying rock called Earth. Those humans had brains. And because of their brains, they developed consciousness—they had the capacity to experience life, with all of its ups and downs. When the brain turned off (that is, when the human died), that marked the end of their capacity for conscious experience. Lights out.

This perspective, often known as *physicalism*, essentially says that reality is physical at its core. All life, and human consciousness, exists because of *random* processes in this physical world.

Basic, right? This worldview is so deeply entrenched in modern thinking that many of us (my former self included) never thought to question it. But the implications are significant. Physicalism often denies that there exist spiritual dimensions, or that there is a benevolent spiritual source ("God") without which we wouldn't even be here. It pooh-poohs notions of an afterlife and the potential reality of psychic abilities. It finds the idea of a "soul" laughable, and it doesn't acknowledge meaning or purpose built into the fabric of reality. Rather, meaning and purpose are arbitrary concepts made up within one's mind—as the physicalist story goes.

Naturally, physicalism tends to dismiss religion—because a "religious" belief is usually outside the realm of verifiability; whereas physicalism is the realm of "science," so we're told. However, physicalism's foundations are based on religion. And that's not an exaggeration. More specifically, physicalism is based on an unverified belief that there was an independent world outside of all consciousness—a world that pre-existed consciousness. Simply put: There was a physical world first, and consciousness came later. That world would still be there if all consciousness were to disappear.

There's a big problem with this concept, however. No one can verify that this world was there before consciousness. And no one can verify that the world would still be there if consciousness disappeared. This is because of one simple truth: consciousness is required to experience something. Without a consciousness to verify that something exists, we're dealing in the realm of inference and faith.

The physicalist worldview *depends* on the existence of a world outside consciousness—one that, by definition, cannot be directly observed or verified—claiming that it led to the consciousness we experience in this moment. Stated another way, physicalism contends that the unknown (a world outside consciousness) creates the known (consciousness). It's like saying, "Grant me this one unscientific and unverifiable miracle, then I'll use science to explain everything else. Just trust me on this one." Dr. Bernardo Kastrup surgically exposes physicalism's philosophical weaknesses in his many works, including *The Idea of the World* (2019). As Kastrup's arguments demonstrate, physicalism is—ironically—its own form of religion because it's based on something that by its very nature can never be directly known or observed or verified.

But things get even worse for physicalism. No one has ever observed consciousness popping out of a brain. All that's been observed are correlations between brain states and states of consciousness. However, correlation doesn't necessarily imply causation. Two things can be related without revealing what is causing what. Consider an alternative: What if the brain is like an antenna that receives and transmits consciousness (the "signal")? Or what if the brain is like a filtering apparatus, or an interface, that shows us a limited sliver of reality through our biological organs? These alternative perspectives don't render the brain meaningless but rather suggest that it plays a different role. The brain might simply be an important organ through which our spiritual, "divine" essence has a human experience via the vessel of the human body. This would truly turn mainstream scientific thinking on its head and bring new possibilities about the meaning of life. At the very least, it evokes more hope and wonder than the grim, superstitious worldview of physicalism.

Moreover, as Brett has suggested, perhaps there exists a universal consciousness (God?) that is the basis of all reality and serves as the source of your very existence. What we consider to be the "physical" world is only *perceived* that way, but actually it's all made of consciousness at its core. Think about it. Whatever we experience in life manifests within our own consciousness. Every single time, without exception.

Kastrup likens this perspective to "whirlpools in a stream": all reality is made of "water" (consciousness), and we feel like we're separate because we experience the world through "whirlpools" (our individuated body and soul essence). Separation is on the one hand an illusion and on the other hand a basic reality in daily living.

There exists a shockingly large body of evidence to support this perspective, which I cover in my first book, *An End to Upside Down Thinking* (2018), among other works. The research comes from credible institutions like the University of Virginia's Division of Perceptual Studies, Princeton University's Engineering Anomalies Research Lab (run by Princeton's former Dean of Engineering, Dr. Robert Jahn), the Institute of Noetic Sciences (where I am a Board member), the US government (as revealed by declassified documents validating that psychic abilities are real), among others. There's even peer-reviewed research, such as Dr. Etzel Cardeña's 2018 paper in *American Psychologist*, the official peer-reviewed academic journal of The American Psychological Association, which shows strong statistical evidence for psychic phenomena; and his follow-up paper published in 2025 in *The International Review of Psychiatry* strengthens the case. But none of this is new. The former president of the American Statistical Association, Dr. Jessica Utts, stated back in 1995: "Using the standards applied to any other area of science, it is concluded that psychic functioning has been well established." Science of this kind marks a paradigm shift in our collective thinking. Arguably it represents a *meta*-paradigm shift: it's the paradigm underlying all other paradigms because it shifts our view about the nature of reality itself. Consequently, those entrenched within physicalism (which represents a large portion of mainstream academia) dismiss this new paradigm and any science that supports it. Hubris is one reason. Imagine being a professor who has built a career around a specific way of thinking, and new science comes along that renders much of that past work incomplete or obsolete. It can be difficult to admit, publicly, that "I was wrong." In some cases, there's even suppression of science. For example, neuroscientist Mario Beauregard, PhD, recalled in a 2022 interview that he was told by the head of a prominent neuroscience institute:

"As long as I'm alive, and I'm controlling [this institute], you'll never do neuroscience studies on spirituality. Never." This dynamic makes the funding environment more challenging for "alternative" consciousness research, as well. But hopefully things are starting to shift. As the 19th century German philosopher Arthur Schopenhauer said: "All truth goes through three stages. First it is ridiculed. Then it is violently opposed. Finally it is accepted as self-evident."

This paradigm shift underscores the philosophy that has guided Brett's work with deep meaning and purpose. Consider, for instance, a phenomenon known as the *near-death experience*, in which a person has an expanded consciousness at a time when the physical body is barely functional. Cardiac arrest is one such example. In some of these cases, the person's consciousness hovers over the body. Yes, you read the correctly—people report that they could see and hear things from outside of their body. Upon being resuscitated, the person is able to accurately describe what was perceived—from a vantage point outside the body. Doctors and family members can sometimes verify that the perception was accurate, and they are able to timestamp when that memory occurred, knowing what the body was doing at that time, and confirming that there should have been no way the brain could have created the memory given the body's state of distress. And even if the person had a fully functional brain, how could their consciousness perceive things accurately from outside the body? Physicalism doesn't have a good answer. The book *The Self Does Not Die: Verified Paranormal Phenomena from Near-Death Experiences* (2023) has accumulated over one-hundred of these documented cases. These are verified memories during near-death experiences, which means they aren't just hallucinations caused by a dying brain.

When researchers study phenomena in this domain, they're forced to consider the radical possibility that maybe consciousness doesn't pop out of our brain as physicalists would like us to believe. Perhaps when the brain is turned off, our consciousness is liberated into some other dimension of reality. It's as if a blindfold is lifted.

In this "other dimension of reality," many near-death experiencers report having a life review: they relive the events of their life in a short amount of time. They somehow become the person that they impacted, feeling the events through that other person's eyes. In my eight-episode podcast series, *Where Is My Mind?* (2019), I interviewed bestselling author Dannion Brinkley who has had multiple near-death experiences and life reviews. He told me that he relived his combat days in Vietnam, feeling the pain of the men he killed through their eyes. He even felt the indirect effects of his actions and told me he experienced the pain of the children who would no longer have a father because he killed the father in combat. So moved by these experiences, Brinkley's worldview shifted instantly. He became much less materialistic and even served as a hospice volunteer. When he had near-death experiences later in life, he then felt what it was like to be the person dying in the hospice bed and felt the love and care that he, himself, was emitting.

In essence, the life review points toward the ubiquitous "Golden Rule" that Brett references. The Golden Rule, from this perspective, isn't merely an abstract concept but instead can be viewed as something built into the fabric of reality itself. As the University of Virginia's Bruce Greyson, MD, writes in his book *After* (2021), the consistency of these accounts points us toward a notion extending beyond morality into the realm of natural law. Perhaps the Golden Rule is shared across spiritual traditions because they're all accessing a universal truth through their own lens.

Furthermore, near-death experiences—and also psychedelics, meditation, breathing, and other avenues that enable spiritually-transformative experiences—reveal to the experiencers that love is a property of reality itself. That is, they claim that love is literally embedded within consciousness as its innate quality. This implies that our own essence, as a whirlpool within the stream, is made of love too. Such a framework positions love not as "an emotion that emerges from chemical and electrical activity in the brain," but instead as something foundational. Compassion is by extension a core human quality that brings us into alignment with natural law.

This is where Brett's message becomes really serious—because he's speaking to a fundamental truth of our existence. If all reality is made of consciousness, and if consciousness is made of love, then what is there to fear? That is Brett's "mic drop."

But most of us don't feel this on a daily basis. Rick Archer, the host of the podcast *Buddha at the Gas Pump*, makes an astute point about why this might occur. He contends that there are many "levels" of reality that can simultaneously coexist. At the highest level, pure consciousness is made of love. And yet in our human lives we operate from the "whirl-pool" perspective—a level of reality that feels divorced from "love as the nature of reality." And so we experience fear.

We're thus presented with a *paradox* that underlies our existence. Our human experience is one of separation. There's a "you" and there's a "me"—even if the highest level of reality is an interconnected stream of consciousness. Because of this apparent separation, fear arises as a natural protective mechanism. Sometimes that fear can be quite rational. For example, fear is a beneficial response for one's survival when encountering a hungry bear in the woods.

Our attention must then turn to discernment. Which fears are legitimate—within this human level of reality—and which are manufactured? Brett aptly points out that we can be "led to fear the wrong things," and he notes that fear can be weaponized as "a central force in our politics, media, minds, and myths."

Discerning truth from falsehood necessarily becomes critical because we want to direct our innate compassion toward appropriate causes. In other words, the mere desire to care for others is not enough. Let's say a woman wants to help her friend who's very sick. She then recommends a supplement because an "expert" said it can alleviate symptoms, only to find that when her friend took her advice and consumed the supplement, her symptoms got worse. In that case, compassion wasn't sufficient because it lacked the discernment required to provide proper direction. Her "compassion" was in fact *harmful*.

More generally, things that sound good don't always result in positive change. And sometimes things that look benevolent on the surface are actually "wolves in sheep's clothing." Deception seems to be baked into our reality. And sometimes matters require nuance and are not so black-and-white.

When examining worldly affairs and societal trends, these

considerations are especially important. Opinions are often formed from assumptions and presuppositions that we take for granted as being true, even if we haven't examined them ourselves. We might believe that smart people who came before us already verified something, when they actually haven't done so with rigor. Similarly, we might form opinions from a limited data set without having seen other data sets that would contradict the established opinions.

For this reason, the spiritual teacher David Hawkins, MD, PhD, often warned that "all knowledge is provisional." We claim to "know" something at a certain point in time, based on the information available at that time. Hawkins wisely counseled that we should remain open to changes in our views when new data enters our sphere. He thus advocated for "radical humility." Humility is one of the characteristics that I so admire about my dear friend Brett, and in this book he does the important service of laying out a series of conversation-starters on hot topics. Hopefully this will lead readers to ponder a wide range of possibilities and data sets, in an ultimate effort to direct their compassion appropriately toward the truth.

The stakes here are quite high, especially when considering the monumental technological changes discussed in this book. As Brett writes: "Whether we enter an unprecedented age of abundance or a preventable tragedy, the choice is ours." It's up to us to prioritize compassion and also to exercise discernment. If we do that, we can ride the technological wave toward the Age of Abundance for All.

In fact, this very concept seems to align with the nature of reality. If, at the foundational level, our identity is a universal consciousness that's the source of all existence and is made of love...then "abundance for all" describes...*the nature of reality*. Brett is thus steering the descent of the higher metaphysical reality into our human world. It's quite a profound directive, but perhaps the most important one imaginable.

And yet the world today is mirroring to us something very different—it's revealing intense problems in seemingly every corner, and they only intensify with technological advancements. It would be irresponsible to just dismiss these issues because one believes that the ultimate reality is made of love. Passivity is not the way forward. From

Brett's optimistic perspective, the right actions will allow us to transcend perceived crises. Other pundits are more pessimistic, however.

Even from a metaphysical level, some cosmologies suggest that we face daunting challenges. For instance, the *Nag Hammadi* scriptures (which cover elements of Gnostic Christianity) describe our world as one governed by malevolent spiritual forces seeking to keep humanity imprisoned and in a state of confusion, distraction, and ignorance, such that we would forget our own divine nature. Even with this ominous origin story, the scriptures view the ultimate reality—the one beyond our world of experience—in a positive way. The scriptures claim that the light will overcome the darkness.

We're all lucky that a man of Brett Hurt's stature has become "activated" such that his immense talents and enormous heart are being directed toward the light. May his devotion to this path be an inspiration to us all and propel us to fill the world with love rather than fear.

Mark Gober
Author of the *"Upside Down"* book series and Board Member at the Institute of Noetic Sciences
Austin, Texas January 21, 2026

About the Author

Brett Hurt is a serial tech entrepreneur and investor. Most recently, he was the CEO and co-founder of data.world, the world's leading data catalog platform, acquired by ServiceNow in 2025.

Previously, Brett founded Bazaarvoice ($1B IPO on NASDAQ) and Coremetrics, which was acquired by IBM for $300M.

Alongside his wife Debra, he co-leads Hurt Family Investments (HFI), backing more than 150 start-ups, fifty venture funds, and a wide range of philanthropic initiatives.

Named Austin's Best CEO (Legacy Award) and an Aspen Institute Henry Crown Fellow, Brett has been building technology since the age of seven.

Love Conquers Fear is Brett's new holding company—an expression of how he's thinking about his highest utility over the next decade, and how he can influence the confluence of technology, consciousness, and courage to help humanity reach the Age of Abundance for All.

Throughout *Love Conquers Fear* you will encounter small phoenix notations, "🕊", that identify information referenced in the book's interactive bibliography. These are active links in the eBook edition and accessible through the QR code above in all printed editions, so that you have the cited authors, references, and resources at your fingertips.

Index

More praise for *Love Conquers Fear*

"*Love Conquers Fear* is a gift to humanity from one of the brightest minds of our time. Hurt has synthesized a staggering amount of wide-ranging research into a hopeful and urgent message our civilization desperately needs at this moment in history."

- Dr. Matthew Hinsley, Executive Director of Austin Classical Guitar and author of *Form & Essence: A Guide to Practicing Truth*

"I met you, Brett, in 2015 as fellow Henry Crown Fellows, and I knew right away you were someone I wanted to know more deeply. You lead with heart, you think with depth, and you move through the world with a rare blend of justice and compassion. Not long after we met, you sent me a copy of *Small Giants*. You had already intuited exactly what I needed as a small business owner who led with heart and went out of your way to offer it. That gesture has always stayed with me because it captures who you are.

Love Conquers Fear carries that same spirit, but on a much broader scale. What's powerful is that this isn't just a book for business leaders. It is for anyone trying to figure out how to show up in a world that feels increasingly complex and uncertain. You make a compelling case that the choice between leading with fear or with love is not abstract. It shows up in our everyday decisions, in our relationships, in our communities, and in how we participate in shaping the future.

This book is a reminder that building a more just and compassionate world is not idealistic. It is practical, necessary work. Whether someone runs a company, leads a team, or is simply trying to live with intention and integrity, this book offers a lens that feels both grounding and urgent."

- Elizabeth Faraut, Founder and CEO of LA LOOP

"At a time when the world is consumed by either AI-evangelists or AI doomerism, *Love Conquers Fear* offers a refreshing and vital paradigm shift. Brett Hurt understands that surviving the current civilizational 'bend in the river' we're in with AI and technology requires more than just better code; it requires an awakening, a consciousness we must embed into the decisions we're making daily. By challenging the materialist worldview and advocating for unity, Brett proves that our technological abundance must be guided by conscious awareness. This book is intelligent, thoughtful, and a triumph, proving that our highest frontier is not artificial intelligence, but human consciousness, spirituality, and the power of choice."

> **- Sol Rashidi, CEO of ExecutiveAI, world's first Chief AI Officer, and best-selling author with nine patents**

"The great practical innovation of capitalism was the shift from a fixed and negative sum games being predominant throughout humanity to a positive sum game becoming the norm. It was a marked shift from a world of finitude and fear to one of abundance. What Brett highlights in *Love Conquers Fear* is another marked shift that we have the potential to undertake in today's world, to a mindset of even greater abundance thanks to innovations in technology and spirituality. Perhaps most importantly, it is just that: an opportunity for every individual, and especially business leaders, to take advantage of."

> **- Alexander McCobin, Founder & CEO, Liberty Ventures Network**

"I've known Brett Hurt for a while now, and his new book *Love Conquers Fear* is exactly what you'd expect—wide-ranging, thoughtful, and grounded in both history and what's coming next.

At a time when most conversations about AI lean toward fear, Brett offers something different: a framework for opportunity.

This isn't a narrow, technical take. He walks the reader from first principles to what an "Age of Abundance" could look like—and how we get there. Along the way, he draws from an incredible range of influences. Any book that can move from Thomas Jefferson to Jim Morrison is worth your time.

What stood out most is how he reframes AI—not as a threat, but as a catalyst. A tool that could help leapfrog entire regions, including parts of Africa, from scarcity to something far more dynamic.

He also takes on bigger questions: What happens to capitalism as technology creates massive value at unprecedented speed? How do we rethink systems built for a different era?

And in classic Brett fashion, he shows you how he got there—laying out the books and ideas that shaped his thinking and turning it into a bit of a "choose your own adventure."

Bottom line: If you're trying to make sense of AI, capitalism, and where we're headed, this is a book worth your time."

- **Joseph Kopser, President of Grayline Group and Co-Founder of USTomorrow; co-author of *Catalyst: Leadership and Strategy in a Changing World***

"*Love Conquers Fear* is a beautiful call to action, deeply researched, written in plain language. As AI emerges as the most powerful catalyst ever in humanity's relentless march towards higher consciousness, Brett brings science and spirituality together to inspire us to move from fear toward wisdom. With soul and authenticity. This book is a light."

- **Michael Zeisser, founder of FMZ Ventures**

344

www.ingramcontent.com/pod-product-compliance
Lightning Source LLC
Chambersburg PA
CBHW040804110726
47973CB00031B/283/J